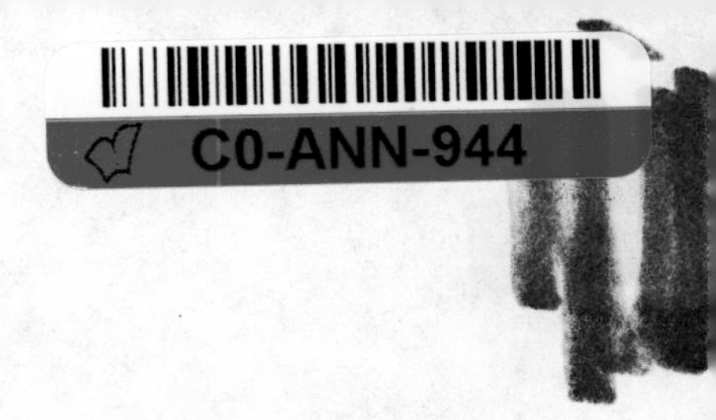

Memoirs of the American Mathematical Society

Number 169

A. J. Lazar and D. C. Taylor

Multipliers of Pedersen's ideal

Published by the

AMERICAN MATHEMATICAL SOCIETY

Providence, Rhode Island

VOLUME 5 · ISSUE 1 · NUMBER 169 (end of issue) · MARCH 1976

ABSTRACT

Let X be a locally compact Hausdorff space and C(X) the algebra of all complex valued continuous functions on X. The purpose of this paper is to make an extensive study of the multipliers of Pedersen's ideal of a C*-algebra, a non-commutative analogue of C(X).

AMS (MOS) subject classifications(1970). Primary 46L05, 46L25; Secondary 46J10.

Key words and phrases. Multipliers, double centralizers, Pedersen's ideal, locally convex algebra.

Library of Congress Cataloging in Publication Data **CIP**

Lazar, Aldo Joram, 1936-
Multipliers of Pedersen's ideal.

(Memoirs of the American Mathematical Society ; no. 169)
"Volume 5, issue 1."
Bibliography: p.
1. C*=algebras. 2. Multipliers (Mathematical analysis) 3. Ideals (Algebra) I. Taylor, Donald Curtis, 1939- joint author. II. Title. III. Series: American Mathematical Society. Memoirs ; no. 169.
QA3.A57 no. 169 [QA326] 510'.8 [512'.55]
ISBN 0-8218-1869-4 75-44302

TABLE OF CONTENTS

MULTIPLIERS OF PEDERSEN'S IDEAL

CHAPTER 1. INTRODUCTION

Let A be a C*-algebra and M(A) its two-sided multiplier (double centralizer) algebra. When A is commutative, that is, $A = C_0(X)$, the algebra of all complex valued continuous functions which vanish at infinity on some locally compact Hausdorff space X, then $M(A) = C_b(X)$, the algebra of all bounded complex valued continuous functions on X. The C*-algebra A together with its multiplier algebra M(A) is the non-commutative analogue of the above relationship between $C_0(X)$ and $C_b(X)$, and is generally considered as the algebraic counterpart of the Stone-Čech compactification of X. This non-commutative generalization of the relationship between $C_0(X)$ and $C_b(X)$ was found useful by Busby in his study of extensions of C*-algebras [5], [6], and by Akemann, Elliot, Pedersen, and Tomiyama in their study of derivations of C*-algebras [1]. This non-commutative analogue suggests that it is possible to obtain a non-commutative generalization of the relationship between $C_{00}(X)$ (sometimes denoted $C_c(X)$), the functions in $C_0(X)$ with compact support, and its multiplier algebra C(X), the algebra of all complex valued continuous functions on X. This can be done by studying the two-sided multipliers of an ideal contained in the C*-algebra A which plays a role similar to that of the ideal $C_{00}(X)$ in $C_0(X)$. Such an ideal was shown to exist by Pedersen.

Indeed in [27], Pedersen showed that every C*-algebra A contains a minimal-dense hereditary ideal K. There, and in subsequent papers ([28], [29], and [30]), Pedersen derived many of the properties of K and established that K plays a role in A similar to that of $C_{00}(X)$ to $C_0(X)$. We refer to this minimal-dense hereditary ideal of A as Pedersen's ideal. The purpose of this paper is to make an extensive study of multipliers of Pedersen's ideal.

Received by the editors April 16, 1975.
This research was supported in part by a grant from the United States-Israel Binational Science Foundation.

Now let $\Gamma(K)$ denote the multipliers of Pedersen's ideal K. In Chapter 2 we include those basic concepts and definitions on multiplier algebras and Pedersen's ideal that are pertinent to our study. In Chapter 3, we study $\Gamma(K)$ under the κ-topology, where the κ-topology is a non-commutative analogue of the compact open topology for $C(X)$. In Chapter 4, we give some examples of $\Gamma(K)$ and in Chapter 5 we develop a comprehensive spectral theory and functional calculus for $\Gamma(K)$. In particular, we prove a spectral mapping theorem and consequently we are able to show that the map $f \to f(x)$ of $C(\sigma_{\Gamma(K)}(x))$ into $\Gamma(K)$ is a *-isomorphism which is continuous under the compact open and κ-topologies. We study the dual of $\Gamma(K)$ under the κ-topology in Chapter 6 and there we characterize the dual of $\Gamma(K)$. In Chapter 7, we study homomorphisms of $\Gamma(K)$ and we are able to prove some extension theorems in the sense of Tietze. We study order and ideals in $\Gamma(K)$ in Chapter 8 and prove a decomposition theorem in the sense of Riesz-Pedersen. In Chapter 9 we study derivations of $\Gamma(K)$. We show that if A is a closed ideal of a W*-algebra or if A is C*-algebra with continuous trace and countable approximate identity, then every derivation of $\Gamma(K)$ is inner. In Chapter 10, we study PCS-algebras. We give several reasonable sufficient conditions for the C*-algebra A to be a PCS-algebra.

We shall be utilizing the following notation throughout this paper. Suppose H is a Hilbert space. Then $B(H)$ denotes the C*-algebra of all bounded linear operators on H, $B_0(H)$ the ideal of $B(H)$ consisting of all compact operators, and $B_{00}(H)$ the ideal of $B_0(H)$ consisting of those operators of finite rank. If $D \subseteq B(H)$, then $D[H]$ denotes the linear span of $\{T(h): T \in D, h \in H\}$. For the locally convex space E, E' will denote the dual of E. If M is a subset of E, then M^0 denotes the polar of M. If C is a subset of some partially ordered linear space, then $C^+ = \{c \in C: c \geq 0\}$. For the element a in the C*-algebra A, [a] shall denote the support of a in the W*-algebra A". Any other unexplained notation may be found in [13] or [35].

CHAPTER 2. PRELIMINARIES

In this section we introduce the reader to basic concepts and definitions of Pedersen's ideal of a C*-algebra, and to the theory of multipliers (double centralizers) of algebras in general. The results of Pedersen's ideal, together with other related work, can be found in [27], [28], [29], [30], [31], and [33]. The results on multipliers and other related work may be found in [3], [5], [21], [22], [36], and [37].

Let A be a C*-algebra, $\tilde{A}$ the C*-algebra obtained by adjoining the identity e to A, and A^+ the positive part of A. A face (or order ideal) is a subcone J of A^+ such that for all x in A^+ the conditions $x \leq y$ and $y \in J$ imply $x \in J$. A face J of A^+ is called invariant if $a^*Ja \subseteq J$ for all $a \in A$. Since the inequality $(a+b)^*y(a+b) \leq 2a^*ya + 2b^*yb$ holds for all a,b,y in $\tilde{A}$ and $\tilde{A}$ is the linear span of its unitary elements, it follows that a face J is invariant if and only if $u^*Ju \subseteq J$ for all unitary elements u in $\tilde{A}$. A face J is called strongly invariant if x^*x in J implies xx^* is in J. Obviously, strong invariance implies invariance. A *-subalgebra B of A is called <u>hereditary</u> (or <u>order related</u>, or <u>facial</u>) if $B^+ \equiv B \cap A^+$ is a face and B is the linear span of B^+.

2.1. <u>Lemma</u> [27, Lemma 1.1, p. 132]. Let J be a face of A^+ and define B as the linear span of elements of J. Then B is a hereditary *-subalgebra of A with $B^+ = J$. Moreover, if J is invariant, then B is a two-sided ideal.

2.2. <u>Theorem</u> [27, Theorem 1.3, p. 134]. Let $\mathcal{J}$ be the collection of all dense invariant faces of A^+ and let K_A^+ denote $\cap \mathcal{J}$. Then K_A^+ is also a dense invariant face of A^+. Consequently, the linear span K_A of K_A^+ is a two-sided, dense, hereditary ideal in A, minimal among all such.

From now on we will refer to K_A as the Pedersen ideal of the C*-algebra A. Sometimes, when A is understood, we will denote the Pedersen ideal of A by K.

The next result is very useful in the computation of Pedersen's ideal of certain concrete C*-algebras.

2.3. Proposition. Let K_{00}^{+} be the set of all positive elements x in A such that there is a positive y in A satisfying $xy = x$. Let J_0 be the set of all x in A^{+} such that

$$x \leq \Sigma_{i=1}^{n} y_i$$

for some $y_1, y_2, \ldots, y_n$ in K_{00}^{+}. Then $J_0 = K_A^{+}$.

Proof. The proof is implicit in [27].

2.4. Example. Let X be a locally compact Hausdorff space and $A = C_0(X)$, the complex valued continuous functions on X that vanish at infinity. Then $K_A = C_{00}(X)$, that is, the functions in $C_0(X)$ with compact support.

2.5. Example. Let H be a complex Hilbert space and $A = B_0(H)$ the C*-algebra of all compact operators on H. Then $K_A = B_{00}(H)$, the operators on H with finite rank. More generally, it follows from [12, Proposition 10, p. 11] and 2.2 that if J is the minimal, non-zero, ideal of some factor B and A is the uniform closure of J, then $K_A = J$. Of course if the underlying Hilbert space is separable, then type I_∞ and type II_∞ factors are the only ones with non-trivial ideals.

2.6. Theorem [29, Proposition 4, p. 267]. If $\{x_i\}$ is a finite set of elements from K_A, then the hereditary C*-algebra generated by them is contained in K_A. Consequently, if $x \in K_A^{+}$, then $x^{1/2} \in K_A^{+}$.

2.7. Theorem [28, Theorem 1.6, p. 66]. Let B be a hereditary C*-subalgebra of K_A. Then the spectrum $\hat{B}$ is homemorphic to $\hat{A} \setminus \text{hull } B$. Moreover, if B is generated by an element a in K, then $\hat{B}$ is homemorphic to $\{\pi \in \hat{A} : \pi(a) \neq 0\}$.

2.8. Theorem [30, Corollary 1.2, p. 73]. If $\{x_n\}$ and $\{y_m\}$ are finite sequences in A such that

$$\Sigma x_n x_n^* \leq \Sigma y_m^* y_m,$$

then there exist $z_{nm} \in A$ such that

$$\Sigma_m z_{nm}^* z_{nm} = x_n^* x_n \text{ and } \Sigma_n z_{nm} z_{nm} \leq y_m y_m^*.$$

2.9. Proposition [29, Corollary 6, p. 268]. If ϕ is a *-homomorphism of A onto the C*-algebra B, then $\phi(K_A) = K_B$.

Now let us suppose that J is an algebra over the field of complex numbers. By a multiplier (double centralizer) of J we mean a pair (S,T) of functions from J to J such that $xS(y) = T(x)y$ for all x,y in J, and we denote the set of all multipliers of J by $\Gamma(J)$. The set $\Gamma(J)$ of multipliers is a vector space under the natural operations of addition and scalar multiplication; furthermore, under multiplication defined by $(S,T)(U,V) = (SU,VT)$, where SU and VT means the composition of the given functions, $\Gamma(J)$ is an algebra. If J is a *-algebra, then an involution can be defined on $\Gamma(J)$ by $(S,T)^* = (T^*,S^*)$, where $S^*(x) = S(x^*)^*$ and $T^*(x) = T(x^*)^*$ for each $x \in J$.

Now throughout the remainder of this section, J will denote a normed *-algebra with approximate identity. By an approximate identity for J, we mean a net $\{e_\lambda\}$ in J, $\|e_\lambda\| \leq 1$, such that $x = \lim e_\lambda x = \lim xe_\lambda$ for all $x \in J$. Now suppose (S,T) is a multiplier of J. If S is bounded, then it is easy to show that T is bounded and $\|S\| = \|T\|$. The multiplier (S,T) is called bounded whenever S and T are bounded, or equivalently, S is bounded. We denote all of the bounded multipliers of J by $\Delta(J)$, which is a Banach *-subalgebra of $\Gamma(J)$ under the norm $\|(S,T)\| = \|S\|$.

The results in the remainder of this section have been proven in a more general setting, but for our purposes a normed *-algebra with approximate identity is general enough.

2.10. <u>Lemma</u> [21, Theorem 2, Theorem 7, Theorem 14, Lemma 1]. For each multiplier (S,T) of J the following statements are true: (i) S and T are linear maps from J to J with closed graphs; (ii) $S(xy) = S(x)y$ and $T(xy) = xT(y)$ for all x,y in J. Consequently, if J is a Banach algebra, then $\Delta(J) = \Gamma(J)$.

When J is a C*-algebra we utilize a special notation for $\Gamma(J)$, that is, $M(J) \equiv \Gamma(J)$.

For each $x \in J$ let (L_x, R_x) be the multiplier of J defined by $L_x(y) = xy$ and $R_x(y) = yx$ for all y in J. Then define the map $\mu_0 : J \to \Delta(J)$ by the formula $\mu_0(x) = (L_x, R_x)$. We will refer to μ_0 as the canonical mapping of J into $\Delta(J)$.

2.11. <u>Proposition</u> [21, Theorem 1, Theorem 3, Theorem 6]. For the canonical mapping μ_0 of J into $\Delta(J)$ the following statements are true: (i) μ_0 is a *-isomorphism of J into $\Delta(J)$ and $\mu_0(J)$ is a two-sided ideal of $\Gamma(J)$; (ii) μ_0 is onto if and only if J has an identity; (iii) if J is commutative, then so is $\Gamma(J)$.

2.12. <u>Theorem</u> [5, 2.11, 3.9]. If J is a C*-algebra, then $M(J)$ is a C*-algebra. Moreover, $M(J)$ is *-isomorphic to the idealizer of J in its bidual, that is, $\{y \in J^{**} : xy, yx \in J \text{ for all } x \in J\}$.

2.13. Let J_1 also be a normed *-algebra with approximate identity, and let ϕ be a *-homomorphism of J onto J_1. Then ϕ can be extended in a natural manner to a *-homomorphism $\overline{\phi}$ of $\Gamma(J)$ into $\Gamma(J_1)$. Here $\overline{\phi}(S,T) = (U,V)$ is defined by

$$U(\phi(x)) = \phi(S(x)) \text{ and } V(\phi(x)) = \phi(T(x))$$

for all $x \in J$. Note $\overline{\phi}(\Delta(J)) \subseteq \Delta(J_1)$.

CHAPTER 3. THE κ-TOPOLOGY

Let A be a C*-algebra and K its Pedersen ideal. In this chapter we shall define a certain topology on $\Gamma(K)$ and investigate some of its properties. As we shall see in Chapter 4, if $A = C_0(X)$ for some locally compact Hausdorff space X, then $\Gamma(K)$ can be identified with $C(X)$, and in this case the topology on $\Gamma(K)$, which we define below, is the familiar compact-open topology. We need first some technical results.

3.1. Lemma. Let $\{x_i\}_{i=1}^n$ be elements of K. For every $\epsilon > 0$ there exist $a \in K^+$ and $\{y_i\}_{i=1}^n$, $\{z_i\}_{i=1}^n$ in K such that

$$\|x_i - y_i\| < \epsilon, \ \|x_i - z_i\| < \epsilon, \text{ and } x_i = ay_i = z_i a, \ i = 1,2,\ldots n.$$

Proof. Let B denote the C*-subalgebra of A generated by $\{x_i\}_{i=1}^n$. It follows from 2.6 that $B \subseteq K$. The assertion is now a consequence of a variant of the Cohen-Hewitt factorization theorem [18, Theorem 2.5, p. 151].

For each element $a \in A$ let $\mathcal{L}_a(\mathcal{R}_{a*})$ be the closed left (respectively, right) ideal generated by $a(a*)$. Denote also by $B_{|a|}$ the C*-subalgebra of A generated by $|a|$, where $|a| = (a*a)^{1/2}$. Clearly, $B_{|a|} \subseteq \mathcal{L}_a \cap \mathcal{R}_{a*}$.

3.2. Lemma. Let a be an element of A. Then $\mathcal{L}_a(\mathcal{R}_{a*})$ is a right (left) $B_{|a|}$-module. Every approximate identity for $B_{|a|}$ is a right (left) approximate identity for $\mathcal{L}_a(\mathcal{R}_{a*})$.

Proof. Without loss of generality, we may assume $\|a\| = 1$. The statement is obvious, since $\mathcal{L}_a$ and $\mathcal{R}_{a*}$ are the closures of $\{xa : x \in A\}$ and $\{a*x : x \in A\}$, respectively, and $(a*a)^{1/n}$ is a right (left) approximate identity for $\mathcal{L}_a(\mathcal{R}_{a*})$.

3.3. Proposition. For each $a \in K$ one has $\mathcal{L}_a, \mathcal{R}_a \subseteq K$.

Proof. It is enough to show that $\mathcal{L}_a \subseteq K$. By 3.1 there are $a_1 \in K$, $a_2 \in K^+$ such that $a = a_1 a_2$. Thus, without loss of generality, we may suppose that $a \in K^+$. Now let $x \in \mathcal{L}_a$. It follows from 3.2 and the Cohen-Hewitt factorization theorem ([18, Theorem 2.5, p. 151]) that $x = yz$ for some $y \in \mathcal{L}_a$ and $z \in B_a$. Since $B_a \subseteq K$ by 2.6 we get $x \in K$ and the proof is complete.

3.4. Proposition. Let $a \in K$ and $(S,T) \in \Gamma(K)$. Then $\mathcal{L}_a(\mathcal{R}_a)$ is invariant under S (respectively, T). Consequently, $S|\mathcal{L}_a$ and $T|\mathcal{R}_a$ are bounded linear operators on $\mathcal{L}_a$ and $\mathcal{R}_a$, respectively.

Proof. Let $\{e_\lambda\}$ be an approximate identity for A contained in K and let $x \in \mathcal{L}_a$. Then $S(x) = \lim_\lambda e_\lambda S(x) = \lim_\lambda T(e_\lambda)x$ is in $\mathcal{L}_a$. Hence $\mathcal{L}_a$ is invariant under S. Similarly, $\mathcal{R}_a$ is invariant under T. The fact that $S|\mathcal{L}_a$ and $T|\mathcal{R}_a$ are bounded operators follows from 2.10 and the closed graph theorem.

Obviously, every bounded multiplier of K can be uniquely extended to a multiplier of A. The converse is also true, that is, if $(S,T) \in M(A)$, then $(S_{|K}, T_{|K}) \in \Gamma(K)$. Indeed, for $a \in K^+$ we have by 2.10

$$S(a) = S(a^{1/2})a^{1/2}, \quad T(a) = a^{1/2}T(a^{1/2}),$$

thus K is invariant under S and T by 2.6. Obviously, the map $(S,T) \to (S_{|K}, T_{|K})$ is an isometric *-isomorphism of M(A) onto $\Delta(K)$ and from now on we shall identify K, A and M(A) with the subalgebras of $\Gamma(K)$ corresponding to them by 2.11 and the above remarks.

3.5. Lemma. Let (S,T) be a pair of maps of K into A satisfying $xS(y) = T(x)y$ for all $x,y \in K$. Then $(S,T) \in \Gamma(K)$.

Proof. We have to show that $S(x), T(x) \in K$ for each $x \in K$. Let $\{e_\lambda\}$ be an approximate identity for A contained in K. Then

$$S(x) = \lim_\lambda e_\lambda S(x) = \lim_\lambda T(e_\lambda)x \in \mathcal{L}_x$$

which is contained in K by 3.3. One shows that $T(x) \in K$ in the same way.

3.6. For each $a \in K$ consider the following semi-norms on $\Gamma(K)$:

$$\lambda_a((S,T)) = \|S(a)\|, \ \rho_a((S,T)) = \|T(a)\|.$$

We define the κ-topology on $\Gamma(K)$ to be that locally convex topology generated by the semi-norms $\{\lambda_a\}_{a\in K}$, $\{\rho_a\}_{a\in K}$.

3.7. __Lemma.__ Let $(S,T) \in \Gamma(K)$. Then S and T are κ-continuous operators on K.

__Proof.__ Let $\{x_\alpha\}$ be a net in K which κ-converges to some x in K, that is, $\|x_\alpha y - xy\| \to 0$ and $\|yx_\alpha - yx\| \to 0$ for each $y \in K$. Since $yS(x_\alpha) = T(y)x_\alpha$ and $S(x_\alpha)y = S(x_\alpha y)$, it clearly follows by virtue of our hypothesis and 3.4 that $\|yS(x_\alpha) - yS(x)\| \to 0$ and $\|S(x_\alpha)y - S(x)y\| \to 0$. The proof of the assertion for T is similar.

3.8. __Theorem.__ $\Gamma(K)$ is a locally convex complete topological algebra under the κ-topology.

__Proof.__ We have to show that the map $((S,T),(U,V)) \to (SU,VT)$ of $\Gamma(K) \times \Gamma(K)$ into $\Gamma(K)$ is separately κ-continuous. Let $\{(S_\alpha,T_\alpha)\}$ be a net in $\Gamma(K)$ which κ-converges to (S,T). Let x be any element of K. Clearly $S_\alpha(U(x)) \to S(U(x))$ in the norm topology of K. By 3.1, $x = yz$ for some $y,z \in K$. Thus $VT_\alpha(x) = yV(T_\alpha(z))$. Hence, $VT_\alpha(x)$ converges uniformly to $VT(x)$, since V is κ-continuous by 3.7. This means that $(S_\alpha,T_\alpha)(U,V)$ converges in the κ-topology to $(S,T)(U,V)$. Similarly we can show that $(U,V)(S_\alpha,T_\alpha)$ converges to $(U,V)(S,T)$ in the κ-topology.

To show completeness, let $\{(S_\alpha,T_\alpha)\}$ be a κ-Cauchy net in $\Gamma(K)$. This means that for each $x \in K$ the nets $\{S_\alpha(x)\}$, $\{T_\alpha(x)\}$ are uniformly Cauchy in K. Let

$$S(x) = \lim_\alpha S_\alpha(x), \; T(x) = \lim_\alpha T_\alpha(x).$$

Then S and T are maps of K into A which clearly satisfy $xS(y) = T(x)y$ for every $x,y \in K$. By 3.5, (S,T) is a multiplier of K and clearly (S,T) is the κ-limit of $\{(S_\alpha,T_\alpha)\}$.

As it will be seen in the next chapter, the multiplication in $\Gamma(K)$ need not be jointly κ-continuous.

3.9. Proposition. K is κ-dense in $\Gamma(K)$.

Proof. Let $\{e_\lambda\}$ be an approximate identity for A contained in K. If $(S,T) \in \Gamma(K)$, then by 2.10 and 3.4 $\{S(e_\lambda)\}$ converges to (S,T) in the κ-topology.

3.10. Proposition. The involution is a κ-continuous operation on $\Gamma(K)$.

Proof. If $\lim_\alpha (S_\alpha,T_\alpha) = (S,T)$ in the κ-topology, then

$$\lim_\alpha T^*_\alpha(x) = \lim_\alpha T_\alpha(x^*)^* = T^*(x) \text{ and } \lim_\alpha S^*_\alpha(x) = \lim_\alpha S_\alpha(x^*)^* = S^*(x)$$

for every $x \in K$, thus $\kappa\text{-}\lim_\alpha(S_\alpha,T_\alpha)^* = (S,T)^*$.

3.11. Proposition. Let ϕ be a *-homomorphism of the C*-algebra A onto the C*-algebra B. The extension $\overline{\phi} = \Gamma(K_A) \to \Gamma(K_B)$ given by 2.9 and 2.13 is continuous for the κ-topologies of $\Gamma(K_A)$ and $\Gamma(K_B)$.

Proof. Let $\kappa\text{-}\lim_\alpha(S_\alpha,T_\alpha) = (S,T)$ in $\Gamma(K_A)$ and

$$\overline{\phi}((S_\alpha,T_\alpha)) = (U_\alpha,V_\alpha), \; \overline{\phi}((S,T)) = (U,V).$$

For every $x \in K_A$ we have

$$U(\phi(x)) = \phi(S(x)) = \lim_\alpha \phi(S_\alpha(x)) = \lim_\alpha U_\alpha(\phi(x)) \text{ and}$$

$$V(\phi(x)) = \phi(T(x)) = \lim_\alpha \phi(T_\alpha(x)) = \lim_\alpha V_\alpha(\phi(x)).$$

Thus

$$\kappa\text{-}\lim_\alpha \overline{\phi}((S_\alpha, T_\alpha)) = \overline{\phi}((S,T)).$$

3.12. A typical neighborhood of the origin in the κ-topology is

$$\{(S,T) \in \Gamma(K) : \|S(x_i)\| \leq 1,\ \|T(y_j)\| \leq 1,$$

$$1 \leq i \leq n,\ 1 \leq j \leq m,$$

$$\{x_i\}_{i=1}^n \subset K,\ \{y_j\}_{j=1}^m \subset K\}.$$

For future use we need a smaller base of neighborhoods.

3.13. Proposition. The collection of sets

$$V_a = \{(S,T) \in \Gamma(K) : \|S(a)\| \leq 1,\ \|T(a)\| \leq 1\}$$

for $a \in K^+$ is a neighborhood base at the origin for the κ-topology.

Proof. Clearly V_a is a neighborhood of the origin. Let

$$U = \{(S,T) \in \Gamma(K) : \|S(x_i)\| \leq 1,\ \|T(y_j)\| \leq 1,$$

$$1 \leq i \leq n,\ 1 \leq j \leq m\}$$

for some

$$x_1, x_2, \ldots, x_n,\ y_1, y_2, \ldots, y_m$$

in K.

By 3.1 there are $a_1 \in K^+$ and $\{u_i\}_{i=1}^n$, $\{v_j\}_{j=1}^m$ in K such that

$$x_i = a_1 u_i, \ 1 \leq i \leq n, \text{ and } y_j = v_j a_1, \ 1 \leq j \leq m.$$

Denote

$$M = \max\{\|u_i\|, \|v_j\| : 1 \leq i \leq n, \ 1 \leq j \leq m\}$$

and $a = Ma_1$. Let $(S,T) \in V_a$. Then

$$\|S(x_i)\| = \|S(a_1)u_i\| \leq \|S(a_1)\|\|u_i\| \leq 1$$

and

$$\|T(y_j)\| = \|v_j T(a_1)\| \leq \|v_j\|\|T(a_1)\| \leq 1$$

for $1 \leq i \leq n$, $1 \leq j \leq m$, so $(S,T) \in U$. Hence $V_a \subseteq U$ and the proof is complete.

We conclude this chapter with a problem: is it possible to characterize the κ-compact subsets of $\Gamma(K)$ in an analogous manner to the characterization given by the Arzela-Ascoli theorem for the commutative case?

CHAPTER 4. EXAMPLES

We shall present in this chapter several examples of $\Gamma(K)$ for some C*-algebras. In some instances we shall also identify the κ-topology of $\Gamma(K)$.

4.1. Let X be a locally compace Hausdorff space and $A = C_0(X)$. Then $K_A = C_{00}(X)$ - the ideal of all functions in A with compact support [27, p. 134]. It has been proved in [22, p. 605] that in this case $\Gamma(K_A) = C(X)$. As stated in Chapter 3, the κ-topology of C(X) is the topology of uniform convergence on compact subsets of X. Indeed, let $\{f_\alpha\}$ be a net in C(X) which κ-converges to $f \in C(X)$. Let F be a compact subset of X and ϕ a function in $C_{00}(X)$ such that $\phi_{|F} = 1$. Since $\|(f - f_\alpha)\phi\| \to 0$, we get that $\{f_\alpha\}$ converges to f, uniformly on F. The proof of the converse is just as simple as the above proof.

4.2. Let H be a Hilbert space and $A = B_0(H)$. Then, as remarked in [27, p. 134], $K_A = B_{00}(H)$ - the two-sided ideal of all finite rank operators. We claim that $\Gamma(K_A)$ can be identified with B(H). Moreover, a net $\{T_\alpha\}$ in B(H) κ-converges to $T \in B(H)$ if and only if $\{T_\alpha\}$, $\{T_\alpha^*\}$ converge to T, T* respectively in the strong operator topology. Indeed, it is obvious that B(H) can be identified with a *-subalgebra of $\Gamma(K_A)$ containing K_A. We have to show that B(H) is κ-complete. Let $\{T_\alpha\}$ be a κ-Cauchy net in B(H). Then the nets $\{T_\alpha T\}$, $\{TT_\alpha\}$ are norm-Cauchy for every $T \in B_{00}(H)$. Thus $\{T_\alpha\}$ and $\{T_\alpha^*\}$ are strongly Cauchy nets. For each $h \in H$, let $Th = \lim_\alpha T_\alpha h$, $T'h = \lim_\alpha T_\alpha^* h$. Then T,T' are linear maps of H into H and $\langle Th_1, h_2\rangle = \langle h_1, T'h_2\rangle$ for every $h_1, h_2 \in H$. It follows easily from this that T,T' have closed graphs. Thus $T, T' \in B(H)$ and $T' = T^*$. It is also readily seen that T is the κ-limit of $\{T_\alpha\}$ so B(H) is complete. The assertion about the κ-topology is immediate.

It is well-known that the multiplication in $B(H)$ is not jointly continuous for the *-strong operator topology if H is infinite dimensional. Thus the product in $\Gamma(K)$ is not, in general, jointly κ-continuous.

For $A = B_0(H)$ we have $\Delta(K_A) = M(A) = \Gamma(K)$. This is due to the fact that $K_A[H] = H$ as we shall see below.

4.3. <u>Proposition</u>. Let A be a C*-algebra of operators on a Hilbert space H such that $A[H] = H$ and let K be its Pedersen ideal. Then $\Gamma(K)$ is *-isomorphic with the *-algebra of all linear operators a from $H' \equiv K[H]$ to H' which have the property that $xa + ay$ is bounded for every $x,y \in K$ and the unique extension of $xa + ay$ to $\overline{H'} = H$ is in K.

The set of all linear operators a on H' with the above property forms an algebra under the usual operations. It is a *-algebra if the adjoint of a is taken to be the restriction of a^* (the usual Hilbert space adjoint for a not necessarily bounded linear operator) to H'. Indeed, if $x \in K$, $\xi \in H$ and $\eta \in H'$, then $\langle a\eta, x\xi\rangle = \langle x^*a\eta, \xi\rangle = \langle y\eta, \xi\rangle = \langle \eta, y^*\xi\rangle$ for some $y \in K$. Thus $\langle a\eta, x\xi\rangle$ is continuous in η which means that a^* is defined on H' and $a^*(x\xi) = y^*\xi \in H'$.

<u>Proof</u>. Let $u = (S,T) \in \Gamma(K)$ and define $\hat{u}: H' \to H'$ by

$$\hat{u}\left(\Sigma_{i=1}^{n} x_i\xi_i\right) = \Sigma_{i=1}^{n} S(x_i)\xi_i$$

for $x_i \in K$, $\xi_i \in H$, $1 \leq i \leq n$. Then $\hat{u}$ is a well defined linear operator. Indeed, if

$$\Sigma_{i=1}^{n} x_i\xi_i = 0$$

and $\{e_\alpha\}$ is an approximate identity of A contained in K, then we get

$$\Sigma_{i=1}^{n} S(x_i)\xi_i = \lim_\alpha \Sigma_{i=1}^{n} e_\alpha S(x_i)\xi_i = \lim_\alpha \Sigma_{i=1}^{n} T(e_\alpha)x_i\xi_i = 0.$$

Clearly, for $x \in K$, $x\hat{u}$ and $\hat{u}x$ are the restrictions to H' of xu, $ux \in K$ respectively; hence, $\hat{u}$ belongs to the *-algebra of operators on H' considered above. It is routine to check that $u \to \hat{u}$ is a *-isomorphism of $\Gamma(K)$ into this algebra. On the other hand, if a is an element of this *-algebra, define $S(x)$ and $T(x)$ for $x \in K$ to be the extensions to H of ax and xa respectively. Then $u = (S,T) \in \Gamma(K)$ and $\hat{u} = a$, so the above *-isomorphism is onto.

4.4. Proposition. Let A be a C*-algebra of operators on a Hilbert space H such that $K_A[H] = H$. Then $\Delta(K_A) = M(A) = \Gamma(K_A)$.

Proof. The first equality is always true, as remarked in the preceding chapter. $\Delta(K_A) = \Gamma(K_A)$ follows immediately from 4.3 and its proof (for each $a \in \Gamma(K_A)$, a^* is everywhere defined).

4.5. Proposition. Let A be a C*-algebra and let ϕ be a pure state of A. Then ϕ admits a unique extension to a κ-continuous linear functional on $\Gamma(K_A)$.

Proof. Let π be the irreducible *-representation of A defined by ϕ; H the underlying Hilbert space, and $\xi \in H$ the corresponding cyclic vector. Let $B = \pi(A)$. Then $K_B[H] = H$. Indeed, let $\eta \in H$ and choose $a \in K_A$ so that $\eta(a)\xi \neq 0$. By [13, 2.8.3] there is $x \in A$ so that $\eta = \pi(xa)\xi$, and by 2.9 $\pi(xa) \in K_B$. From 4.4 we infer that $\Gamma(K_B)$ is a *-algebra of bounded operators on H. Let $\overline{\pi}:\Gamma(K_A) \to \Gamma(K_B)$ be the extension of π mentioned in 3.11 and define a linear functional $\overline{\phi}$ on $\Gamma(K_A)$ by $\overline{\phi}(x) = \langle\overline{\pi}(x)\xi,\xi\rangle$. Clearly $\overline{\phi}$ extends ϕ. This extension of ϕ is κ-continuous on $\Gamma(K_A)$ because $\xi = \pi(b)\xi_1$ for some $b \in K_A$, $\xi_1 \in H$. The uniqueness assertion is a consequence of 3.9.

4.6. In view of 4.3, the following problem seems of interest: let $T_1, T_2, \ldots, T_n$ be closed linear operators defined on a dense subspace of a Hilbert space H. Under what condition must there exist a C*-algebra A of operators on H such that $T_i \in \Gamma(K_A)$, $1 \leq i \leq n$?

4.7. Let $\{A_i\}_{i\in I}$ be a family of C*-algebras and A its restricted sum. That is $x = (x_i) \in A$ if and only if for each $\epsilon > 0$ the set $\{i \in I: \|x_i\| > \epsilon\}$ is finite. With pointwise operations and $\|x\| = \sup\|x_i\|$, A is a C*-algebra. Let K_i be the Pedersen ideal of A_i. One sees immediately from the definition of K_A that $x = (x_i) \in A$ belongs to K_A if and only if $x_i = 0$ except for finitely many indices and $x_i \in K_i$ for each $i \in I$. By using the natural *-homomorphism of A onto A_i, one easily recognizes that $\Gamma(K_A)$ can be identified with the sum of the *-algebras $\Gamma(K_i)$, $i \in I$.

CHAPTER 5. SPECTRAL THEORY AND A FUNCTIONAL CALCULUS

Throughout this section, A will denote a C*-algebra, K_A its Pedersen ideal (or simply K if A is understood), and $\Gamma(K)$ the multipliers of K. As in Section 3, we will view K, A and $\Delta(K)$ as subalgebras of $\Gamma(K)$.

5.1. For each $a \in K$, let $\mathcal{L}_a$ and $\mathcal{R}_{a^*}$ denote the closed left and right ideals of A generated by a and a*, respectively. Note that by 3.3, $\mathcal{L}_a$ and $\mathcal{R}_{a^*}$ are subsets of K. Now let M_a denote the set of all pairs (U,V) that satisfy the following: (i) U and V are linear operators on $\mathcal{L}_a$ and $\mathcal{R}_{a^*}$, respectively; (ii) $xU(y) = V(x)y$ for each $x \in \mathcal{R}_{a^*}$ and $y \in \mathcal{L}_a$.

Let (S,T) and (U,V) belong to M_a and let α be a complex number. Then it is clear that (S+U,T+V), $(\alpha U,\alpha V)$, and (SU,VT) belong to M_a. Moreover, if we define S* on $\mathcal{R}_{a^*}$ be the formula $S^*(x) = S(x^*)^*$, and similarly define T* on $\mathcal{L}_a$, then (T*,S*) belongs to M_a. Consequently, M_a is a *-algebra when provided with the following operations:
(i) (S,T) + (U,V) = (S+U,T+V); (ii) $\alpha(S,T) = (\alpha S,\alpha T)$;
(iii) (S,T)(U,V) = (SU,VT); (iv) (S,T)* = (T*,S*).

5.2. <u>Lemma</u>. If $(S,T) \in M_a$, then S and T are bounded and $\|S\|^2 = \|S^*\|^2 = \|T\|^2 = \|T^*S\|$. Consequently, the *-algebra M_a when provided with the norm $\|(S,T)\| = \|S\|$ is a C*-algebra with identity.

<u>Proof</u>. Let $(S,T) \in M_a$. Let $\{x_n\}$ be a sequence in $\mathcal{L}_a$ that converges to x and suppose the sequence $\{S(x_n)\}$ converges to y. Then

$$\|S(x)-y\|^2 = \|(S(x)-y)^*(S(x)-y)\|$$

$$= \lim_n \|(S(x)-y)^*(S(x)-S(x_n))\|$$

$$= \lim_n \| T((S(x)-y)^*)(x-x_n)\|$$

$$= 0.$$

Hence by the closed graph theorem S is bounded. Similarly, T is bounded. Now

$$\|S^*\| = \sup\{\|S^*(y)\| : y \in \mathcal{R}_{a^*}, \|y\| = 1\}$$

$$= \sup\{\|S(x)\| : x \in \mathcal{L}_a, \|x\| = 1\}$$

$$= \|S\|.$$

Similarly, $\|T^*\| = \|T\|$. By using these facts we get

$$\|S\|^2 = \sup\{\|S(x)^*S(x)\| : x \in \mathcal{L}_a, \|x\| = 1\}$$

$$= \sup\{\|T(S(x)^*)x\| : x \in \mathcal{L}_a, \|x\| = 1\}$$

$$= \sup\{\|x^*T^*S(x)\| : x \in \mathcal{L}_a, \|x\| = 1\}$$

$$\leq \|T^*S\| \leq \|T\|\|S\|.$$

Hence, $\|S\| \leq \|T\|$, and similarly, $\|T\| \leq \|S\|$. The above inequality becomes

$$\|S\|^2 \leq \|T^*S\| \leq \|S\|^2.$$

The remainder of the proof is routine.

5.3. Definition. Let $a \in K$. For each multiplier (S,T) of K, define $\lambda_a(S) = S_{|\mathcal{L}_a}, \rho_a(T) = T_{|\mathcal{R}_{a^*}}$, and $\phi_a((S,T)) = (\lambda_a(S), \rho_a(T))$. By virtue of 3.4, $\lambda_a(S)$ and $\rho_a(T)$ are well defined bounded linear operators on $\mathcal{L}_a$ and $\mathcal{R}_{a^*}$, respectively, hence it is clear that ϕ_a is a *-homomorphism of $\Gamma(K)$ into M_a. The map ϕ_a will be referred to as the canonical representation of $\Gamma(K)$ generated by a.

5.4. Lemma. The C*-algebra $M_a \supseteq \Delta(K_{\phi_a(A)}) = \Gamma(K_{\phi_a(A)})$. Moreover, if A is separable, then $\phi_a(\Gamma(K_A)) = \Gamma(K_{\phi_a(A)})$.

Proof. Let $\{x_\alpha\}$ be a net in K_A for which $\{\phi_a((L_{x_\alpha}, R_{x_\alpha}))\}$ is $\kappa_{\phi_a(A)}$-Cauchy. Let $b \in \mathfrak{L}_a$. By 3.2 the Cohen-Hewitt factorization theorem applies, so b = cd, where c and d belong to $\mathfrak{L}_a$. Due to the facts that $L_{x_\alpha}(b) = L_{x_\alpha}L_c(d)$ and $\phi_a(K) = K_{\phi_a(A)}$ by 2.9, we easily see that $L_{x_\alpha}(b)$ is uniformly Cauchy in $\mathfrak{L}_a$. Define $L(b) = \lim L_{x_\alpha}(b)$. In a similar fashion a map $R: \mathfrak{R}_{a*} \to \mathfrak{R}_{a*}$ can be defined. The pair (L,R) clearly satisfies all the conditions of an element in M_a. Hence the first part of our conclusion follows from 3.9 and 5.2. The second part follows from [3, Theorem 9.2, p. 290] and our proof is complete.

Question. Under what conditions must $M_a = \Delta(K_{\phi_a(A)})$ for each $a \in A$?

5.5. Lemma. Suppose $(U,V) \in M_a$. If U and V are one-to-one and onto maps of $\mathfrak{L}_a$ and $\mathfrak{R}_{a*}$, respectively, then (U^{-1}, V^{-1}) belongs to M_a. Moreover, a multiplier (S,T) of K is regular in $\Gamma(K)$ if and only if S and T are one-to-one and onto maps of K.

Proof. Let $x \in \mathfrak{R}_{a*}$ and $y \in \mathfrak{L}_a$ and suppose (U,V) satisfies the hypothesis. Clearly, U^{-1} and V^{-1} are bounded linear operators of $\mathfrak{L}_a$ and $\mathfrak{R}_{a*}$, respectively. Since U is an onto map, there is a $z \in \mathfrak{L}_a$ such that $U(z) = y$. Thus $V^{-1}(x)y = V^{-1}(x)U(z) = V(V^{-1}(x))z = xU^{-1}(y)$. Hence $(U^{-1}, V^{-1}) \in M_a$. The proof of the second assertion is similar.

5.6. Theorem. Suppose $\{e_\lambda\}_{\lambda \in I}$ is a positive approximate identity for A contained in K. Fhen for each x in $\Gamma(K)$ the following statements are equivalent: (i) x is regular in $\Gamma(K)$; (ii) for each $a \in K$, $\phi_a(x)$ is regular in M_a; (iii) for each $\lambda \in I$, $\phi_{e_\lambda}(x)$ is regular in M_{e_λ}.

Proof. It is clear that (i) implies (ii) and (ii) implies (iii), so assume x = (S,T) is a multiplier of K for which (iii) holds. To prove (i) it will suffice to show, by virtue of 5.5, that the maps

$S:K \to K$ and $T:K \to K$ are one-to-one and onto. Let $a \in K$ such that $S(a) = 0$. Then by 2.10, $S(ae_\lambda) = S(a)e_\lambda = 0$ for all $\lambda \in I$. Since $ae_\lambda \in \mathcal{L}_{e_\lambda}$ and $\lambda_{e_\lambda}(S) = S_{|\mathcal{L}_{e_\lambda}}$ is one-to-one, it follows that $ae_\lambda = 0$ for each $\lambda \in I$. Hence $a = 0$ and S is one-to-one. Similarly, we can show that T is one-to-one. To prove that S and T are onto, it will suffice to show that $K^+ \subseteq S(K)$.

Let $a \in K^+$. The smallest invariant face J containing $\{e_\lambda\}_{\lambda \in I}$ is the set of all elements x for which there are elements $\{e_{\lambda_i}\}_{i=1}^n$, unitary elements $\{u_i\}_{i=1}^n$ in $\tilde{A}$, and positive scalars $\{\alpha_i\}_{i=1}^n$ such that

$$0 \le x \le \Sigma_{i=1}^n \alpha_i u_i^* e_{\lambda_i} u_i.$$

Here $\tilde{A}$ denotes the C*-algebra formed by adjoining the identity to A. Clearly, $J \subset K^+$. Since J is invariant, $b^{1/2}e_\lambda b^{1/2} \in J$ for each $b \in A^+$. Thus J is dense in A^+ which means that $K^+ = J$ by virtue of 2.2. It follows that

$$0 \le a \le \Sigma_{i=1}^n \alpha_i u_i^* e_{\lambda_i} u_i,$$

where α_i, u_i, and e_{λ_i} are described above. By 2.8, there are elements $\{z_i\}_{i=1}^n$ in A such that

$$a = \Sigma_{i=1}^n z_i^* z_i \text{ and } z_i z_i^* \le \alpha_i e_{\lambda_i}, \ 1 \le i \le n.$$

But $z_i z_i^* \le \alpha_i e_{\lambda_i}$ implies

$$\|(1 - e_{\lambda_i}^{1/p})z_i\|^2 = \|(1 - e_{\lambda_i}^{1/p})z_i z_i^*(1 - e_{\lambda_i}^{1/p})\|$$

$$\le \alpha_i \|(1 - e_{\lambda_i}^{1/p})e_{\lambda_i}(1 - e_{\lambda_i}^{1/p})\|.$$

Hence

$$\lim_{p\to\infty} e_{\lambda_i}^{1/p} z_i = z_i .$$

It follows from 3.2 that $z_i \in \mathcal{R}_{e_{\lambda_i}}$ or, equivalently, $z_i^* \in \mathcal{L}_{e_{\lambda_i}}$.

Since $S(\mathcal{L}_{e_\lambda}) = \mathcal{L}_{e_\lambda}$, there are $y_i \in \mathcal{L}_{e_{\lambda_i}}$ such that $S(y_i) = z_i^*$.

Thus

$$a = \Sigma_{i=1}^{n} z_i^* z_i = \Sigma_{i=1}^{n} S(y_i)z_i = \Sigma_{i=1}^{n} S(y_i z_i),$$

hence $a \in S(K)$.

Since S and T are one-to-one and onto, it follows from 5.5 that (S,T) is regular in $\Gamma(K)$ and our proof is complete.

5.7. We had hoped that for a multiplier $x = (S,T)$ of K the following statement would be true: x is regular in $\Gamma(K)$ if and only if $\overline{\pi}(x)$ is regular in $\Gamma(K_{\pi(A)})$ for each irreducible representation π of A. Unfortunately, this is not the case as the following example demonstrates.

Let A be the C*-algebra of all bounded sequences of 2×2 matrices

$$\left\{\begin{pmatrix} a_n & b_n \\ c_n & d_n \end{pmatrix}\right\}$$

such that $\lim b_n = \lim c_n = \lim d_n = 0$. This algebra can be considered as a continuous field of elementary C*-algebra on βN and by [13, 10.4.4, p. 196] $\hat{A} = \beta N$. Since βN is compact, $\Delta(K_A) = \Gamma(K_A)$ (see 10.8, Chapter 10). Clearly, the sequence

$$x = \left\{\begin{pmatrix} 1 & 0 \\ 0 & 1/n \end{pmatrix}\right\}$$

is a multiplier of A and for every $\pi \in A$, $\pi(x)$ is regular. But there is no $y \in \Gamma(K_A) = \Delta(K_A)$ such that $\pi(y) = \pi(x)^{-1}$ for each $\pi \in A$. Indeed, the sequence

$$\left\{\begin{pmatrix} 1 & 0 \\ 0 & n \end{pmatrix}\right\}$$

is unbounded.

Let B be an algebra with identity e and let $x \in B$. The spectrum of x with respect to B, denoted by $\sigma_B(x)$, is defined to be the set of all complex numbers λ for which $\lambda e - x$ is singular in B.

5.8. Corollary. If $x \in \Gamma(K)$, then

$$\sigma_{\Gamma(K)}(x) = \bigcup_{a \in K} \sigma_{M_a}(\phi_a(x)) = \bigcup_{\lambda \in I} \sigma_{M_{e_\lambda}}(\sigma_{e_\lambda}(x)).$$

5.9. Corollary. For each $x \in \Gamma(K)$, $\sigma_{\Gamma(K)}(x)$ is nonempty.

5.10. Corollary. If A has a countable approximate identity, then for each $x \in \Gamma(K)$, $\sigma_{\Gamma(K)}(x)$ can be expressed as a countable union of compact sets.

5.11. Corollary. If x is a hermitian element of $\Gamma(K)$, then $\sigma_{\Gamma(K)}(x)$ is real.

5.12. Corollary. If B is a maximal commutative self-adjoint subalgebra of $\Gamma(K)$ and $x \in B$, then $\sigma_B(x) = \sigma_{\Gamma(K)}(x)$.

5.13. Let $x = (S,T)$ be a normal multiplier of K and let f be a complex valued continuous function on $\sigma_{\Gamma(K)}(x)$. By virtue of 5.8, f is continuous on each set $\sigma_{M_a}(x)$, $a \in K$. Since $\phi_a(x)$ is normal in M_a, $f(\phi_a(x))$ can be defined in M_a in the usual way. For each $a \in K$, let

$$(f(\lambda_a(S)), f(\rho_a(T)))$$

denote the element in M_a given by $f(\phi_a((S,T)))$. If f_n is a function of the form

$$f_n(\upsilon) = \Sigma_{i,j}^{n} \alpha_{ij}\upsilon^{i}\,\bar{\upsilon}^{j},$$

υ complex, set

$$f_n(x) = \Sigma_{i,j}^{n} \alpha_{ij}x^{i}(x^*)^{j},$$

$$f_n(S) = \Sigma_{i,j}^{n} \alpha_{ij}S^{i}(T^*)^{j},$$

and
$$f_n(T) = \Sigma_{i,j}^{n} \alpha_{ij}(S^*)^{j}T^{j}.$$

Note that

$$f_n(x) = (f_n(S),\ f_n(T)) \text{ and } \phi_a(f_n(x)) = f_n(\phi_a(x)),$$

or equivalently,

$$\lambda_a(f_n(S)) = f_n(\lambda_a(S)) \text{ and } \rho_a(f_n(T)) = f_n(\rho_a(T)).$$

This suggests the following lemma for the maps $f(S):K \to K$ and $f(T):K \to K$ defined by the formulas

$$f(S)(a) = f(\lambda_a(S))(a) \text{ and } f(T)(a^*) = f(\rho_a(T))(a^*),$$

$a \in K$.

5.14. <u>Lemma</u>. The pair $(f(S),\ f(T))$ is a multiplier of K, and, for each

$$a \in K,\ \phi_a((f(S),f(T))) = f(\phi_a((S,T))).$$

<u>Proof</u>. Let u,v be elements in K. By 3.1, there exists an $a \in K^+$ such that $u \in \mathcal{R}_{a*}$ and $v \in \mathcal{L}_a$. By the Stone-Weierstrass theorem, there is a sequence $\{f_n\}$ of continuous functions of the form

$$f_n(\nu) = \Sigma_{i,j}^n \alpha_{ij}\nu^i \bar{\nu}^j,$$

ν complex, that converges uniformly to f on compact sets. It follows from 5.8 that $f_n(\phi_b(x))$ converges uniformly to $f(\phi_b(x))$ for each $b \in K$. Hence, $f_n(\lambda_b(S))$ and $f_n(\rho_b(T))$ converge uniformly to $f(\lambda_b(S))$ and $f(\rho_b(T))$, respectively. Since

$$v \in \mathcal{L}_a,\ f(S)(v) = f(\lambda_v(S))(v) = \lim f_n(\lambda_v(S))(v) = \lim \lambda_v(f_n(S))(v)$$

$$= \lim \lambda_a(f_n(S))(v) = \lim f_n(\lambda_a(S))(v) = f(\lambda_a(S))(v).$$

Similarly, we can show

$$f(T)(u) = f(\rho_a(T))(u).$$

Now

$$u\, f(S)(v) = u\, f(\lambda_a(S))(v) = \lim u\, f_n(\lambda_a(S)(v)) = \lim u\, f_n(S)(v)$$

$$= \lim f_n(T)(u)v = \lim f_n(\rho_a(T))(u)v = f(T)(u)v.$$

Hence $(f(S), f(T))$ is a multiplier of K. It is clear that

$$\phi_a((f(S),f(T)) = f(\phi_a((S,T)))$$

and our proof is complete.

5.15. Definition. Let x be a normal element of $\Gamma(K)$ and f a complex valued continuous function on $\sigma_{\Gamma(K)}(x)$. We define f(x) to be the element in $\Gamma(K)$ given by 5.14.

5.16. Remark. Let x be a normal element in $\Delta(K)$ and f a complex valued continuous function defined on $\sigma_{\Delta(K)}(x)$. Put $f|\sigma_{\Gamma(K)}(x) = f$. This equality makes sense, since

$$\sigma_{\Gamma(K)}(x) \subseteq \sigma_{\Delta(K)}(x).$$

If f(x) is the element in $\Delta(K)$ defined in the usual way and $\tilde{f}(x)$ is the element in $\Gamma(K)$ given by 5.15, then it is straightforward to show that $f(x) = \tilde{f}(x)$. Thus the definition given by 5.15 is consistent with the usual functional calculus for C*-algebras.

5.17. Corollary. Suppose $\{f_\alpha\}$ is a net of complex valued continuous functions defined on $\sigma_{\Gamma(K)}(x)$ that converges to f in the compact open topology. Then $f_\alpha(x)$ converges to f(x) in the κ-topology.

Proof. Let $a \in K^+$. By virtue of 5.8, $f_\alpha(\lambda_a(x))$ converges uniformly to $f(\phi_a(x))$. Therefore, $f_\alpha(\lambda_a(S))$ and $f_\alpha(\rho_a(T))$ converge uniformly to $f(\lambda_a(S))$ and $f(\rho_a(T))$, respectively. Now

$$f_\alpha(S)(a) = \lambda_a(f_\alpha(S))(a) = f_\alpha(\lambda_a(S))(a),$$

so $f_\alpha(S)(a)$ converges to $f(S)(a)$. Similarly, $f_\alpha(T)(a)$ converges to $f(T)(a)$. Hence by 3.6, $f_\alpha(x)$ converges to f(x) in the κ-topology and our proof is complete.

Let B be a C*-algebra, ψ a *-homomorphism of A onto B, $\overline{\psi}$ the induced *-homomorphism of $\Gamma(K_A)$ into $\Gamma(K_B)$. Since

$$\sigma_{\Gamma(K_B)}(\overline{\psi}(x)) \subseteq \sigma_{\Gamma(K_A)}(x),$$

the following result makes sense. The result improves 5.14.

5.18. Corollary. The multiplier $\overline{\psi}(f(x)) = f(\overline{\psi}(x))$.

Proof. In 5.14, it was observed that there exists a sequence $\{f_n\}$ of complex valued continuous functions of the form

$$f_n(\nu) = \Sigma_{i,j}^n \alpha_{ij}\nu^i \bar{\nu}^j ,$$

ν complex, that converges to f in the compact open topology. Hence

$$f_n(x) = \Sigma_{i,j}^n \alpha_{ij}x^i(x^*)^j$$

converges to f(x) in the κ_A-topology. Since we have by 3.11 that $\overline{\psi}$ is continuous under the κ_A, κ_B topologies, we get from 5.17 that

$$\overline{\psi}(f(x)) = \lim \overline{\psi}(f_n(x)) = \lim f_n(\overline{\psi}(x)) = f(\overline{\psi}(x))$$

and our proof is complete.

5.19. Theorem (spectral mapping theorem). Let x be a normal element in $\Gamma(K)$. If f is a complex valued continuous function on $\sigma_{\Gamma(K)}(x)$, then

$$\sigma_{\Gamma(K)}(f(x)) = f(\sigma_{\Gamma(K)}(x)).$$

Proof. By virtue of 5.8 and 5.14,

$$\sigma_{\Gamma(K)}(f(x)) = \bigcup_{a\in K} \sigma_{M_a}(\phi_a(f(x))) = \bigcup_{a\in K} \sigma_{M_a}(f(\phi_a(x)))$$

$$= \bigcup_{a\in K} f(\sigma_{M_a}(\phi_a(x))) = f(\bigcup_{a\in K} \sigma_{M_a}(\phi_a(x)))$$

$$= f(\sigma_{\Gamma(K)}(x)).$$

5.20. Theorem. Let x be a normal element in $\Gamma(K)$, f and g complex valued continuous functions on $\sigma_{\Gamma(K)}(x)$, h a complex valued continuous function on $\sigma_{\Gamma(x)}(g(x))$, and α, β complex numbers. Then the following statements hold:

(i) $\alpha f(x) + \beta g(x) = (\alpha f + \beta g)(x)$;

(ii) $f(x)g(x) = f \cdot g(x)$;

(iii) if f has a power series expansion $f(\nu) = \Sigma_{k=0}^{\infty} \alpha_k \nu^k$ valid in $\sigma_{\Gamma(K)}(x)$, then $f(x) = \Sigma_{k=0}^{\infty} \alpha_k x^k$, where $\Sigma_{k=0}^{\infty} \alpha_k x^k$ is defined to be the limit of the sequence $\{\Sigma_{k=1}^{n} \alpha_k x^k\}$ in the κ-topology.

(iv) if x is hermitian and f is real valued, then $f(x)$ is hermitian;

(v) $F(x) = h(g(x))$, where $F(\nu) = h(g(\nu))$.

Proof. First, note that for elements y, z in $\Gamma(K)$, $y = z$ if and only if $\phi_a(y) = \phi_a(z)$ for each $a \in K$. The proof now follows from 5.14 and the usual functional calculus for C*-algebras.

5.21. Corollary. The map $f \to f(x)$ of $C(\sigma_{\Gamma(K)}(x))$ into $\Gamma(K)$ is a *-isomorphism which is continuous under the compact open and κ-topologies.

5.22. Example. The following example shows that the map $f \to f(x)$ is not bicontinuous in general even if A is commutative. Let X be the union of the real line and a point p in the plane, not on the real line, under the relative topology of the plane. Let $A = C_0(X)$ and $x \in \Gamma(K) = C(X)$ defined by

$$x(t) = \begin{cases} \pi/2 & t = p \\ \arctan t & t \neq p \end{cases}$$

Now let $Y = \{x(t): t$ is a positive integer or $t = p\}$ under the relative topology, which is clearly a compact subset of

$$\sigma_{\Gamma(K)}(x) = (-\pi/2, \pi/2].$$

Consider the following neighborhood of zero in $C(\sigma_{\Gamma(K)}(x))$:

$$U = \{f \in C(\sigma_{\Gamma(K)}(x)): |f(x)| < 1/2, \ s \in Y\}.$$

We claim that there is no basic neighborhood V of zero in C(X) such that

$$V \cap \tau(C(\sigma_{\Gamma(K)}(x)) \subseteq \tau(U),$$

where

$$\tau: C(\sigma_{\Gamma(K)}(x)) \rightarrow C(X)$$

is the *-isomorphism given in 5.21. Indeed let F be a compact subset of X. It is clear that x(F) is a compact subset of $\sigma_{\Gamma(K)}(x)$ such that $x(F) \neq Y$. Let $\epsilon > 0$ and set

$$V = \{g \in C(X): |g(t)| < \epsilon, \ t \in F\}.$$

Define f in $C(\sigma_{\Gamma(K)}(x))$ such that it vanishes on x(F) but takes the value one at some point of $Y \setminus x(F)$. It follows that $f \notin U$, so $\tau(f) \notin \tau(U)$. But

$$\tau(f) \in V \cap \tau(C(\sigma_{\Gamma(K)}(x)).$$

5.23. Corollary. If x is a hermitian element in $\Gamma(K)$, then the following statements are equivalent:

(i) $\sigma_{\Gamma(K)}(x) \geq 0$;

(ii) x is of the form yy^*, where $y \in \Gamma(K)$;

(iii) x is of the form h^2, where h is a hermitian element in $\Gamma(K)$.

Proof. By using 5.8, 5.19, 5.20, the proof becomes standard.

5.24. Definition. We say that an element x in $\Gamma(K)$ is positive, denoted by $x \geq 0$, if it is hermitian and satisfies any of the equivalent conditions of 5.23. We let $\Gamma(K)^+$ denote the set of all positive elements of $\Gamma(K)$.

5.25. Corollary. The algebra $\Gamma(K)$ is the linear span of its positive elements.

Now let us view A as a C*-algebra of operators on the Hilbert space H such that $A[H] = H$. The algebra $\Gamma(K)$ can be viewed as the *-algebra of all operators T acting in the dense subspace $K[H]$ which have the property $xT + Ty$ is bounded on $K[H]$ for each $x,y \in K$ and its unique extension to H belongs to K (see 4.3).

5.26. Corollary. Let $x \in \Gamma(K)$. Then $x \geq 0$ if and only if $\langle x(h),h\rangle \geq 0$ for each $h \in K[H]$.

Proof. By using 5.20, the proof becomes identical to [13, 1.6.7, p. 14].

5.27. Corollary. The set $\Gamma(K)^+$ is a convex cone and $\Gamma(K)^+ \cap (-\Gamma(K)^+) = 0$.

5.28. Corollary. The algebra $\Gamma(K)$ is symmetric.

The relation $x - y \geq 0$ is an order relation on $\Gamma(K)$ compatible with the real vector space structure of $\Gamma(K)$; we denote this relation by $x \geq y$ or $y \leq x$.

5.29. Proposition. Let a,b,x be elements of $\Gamma(K)$. If $a \leq b$, then $x^*ax \leq x^*bx$. Moreover, if a and b are regular and $b \geq a \geq 0$, then $a^{-1} \geq b^{-1}$.

Proof. The proof is similar to [13, 1.6.8, p. 14] by virtue of 5.19 and 5.20.

5.30. *Proposition.* A normal element x in $\Gamma(K)$ belongs to $\Delta(K)$ if and only if $\sigma_{\Gamma(K)}(x)$ is bounded.

Proof. Suppose $x = (S,T)$ is a normal element in $\Gamma(K)$ with $\sigma_{\Gamma(K)}(x)$ bounded. Let $\alpha > 0$ such that $|\lambda| < \alpha$ whenever $\lambda \in \sigma_{\Gamma(K)}(x)$. By virtue of 5.8, $\|\phi_a(x)\| < \alpha$ for each $a \in K$. This means $\|S(a)\| < \alpha\|a\|$ for each $a \in K$. Hence $x \in \Delta(K)$.

5.31. *Example.* If x is not normal it may happen that $\sigma_{\Gamma(K)}(x)$ is bounded yet $x \notin \Delta(K)$. Suppose A is the algebra of all sequences of 2×2 matrices which converge to the zero matrix. The Pedersen ideal K of A is the space of all sequences that vanish except finitely many times, $\Delta(K)$ is the space of all bounded sequences, and $\Gamma(K)$ is the space of all sequences. Let T be a nilpotent matrix of norm 1. Then $x = \{nT\}_{n=1}^{\infty}$ belongs to $\Gamma(K)$, but not to $\Delta(K)$. However, $\sigma_{\Gamma(K)}(x) = \{0\}$.

5.32. *Proposition.* If x is a normal element in $\Delta(K)$, then the closure of $\sigma_{\Gamma(K)}(x)$ is $\sigma_{\Delta(K)}(x)$.

Proof. Clearly, the closure of $\sigma_{\Gamma(K)}(x)$ is contained in $\sigma_{\Delta(K)}(x)$. Let $\alpha_0 \in \sigma_{\Delta(K)}(x)$. If α_0 is not in the closure of $\sigma_{\Gamma(K)}(x)$, then $f(\alpha) \equiv 1/(\alpha-\alpha_0)$ is a bounded continuous function on $\sigma_{\Gamma(K)}(x)$. Since x is normal, it follows from 5.19 and 5.30 that $f(x)$ belongs to $\Delta(K)$. Thus, by virtue of 5.20 (ii), $\alpha_0 \notin \sigma_{\Gamma(K)}(x)$ which is a contradiction.

5.33. *Question.* Is 5.32 valid without the assumption that x is normal?

5.34. *Proposition.* If x is a normal element in A, then

$$\sigma_{\Delta(K)}(x) \setminus \{0\} \subseteq \sigma_{\Gamma(K)}(x) \subseteq \sigma_{\Delta(K)}(x).$$

Proof. The last inclusion is obvious. Now let

$$\alpha \in \sigma_{\Delta(K)}(x) \setminus \{0\}.$$

Then $\alpha e - x$ is either contained in a left ideal of $\Delta(K)$ or a right ideal. Assume, without loss of generality, that $\alpha e - x$ belongs to a left ideal of $\Delta(K)$. It follows that there is a maximal left ideal of $\Delta(K)$ which contains $\alpha e - x$. Hence there is a pure state f of $\Delta(K)$ such that

$$f((\alpha e - x)^*(\alpha e - x)) = 0,$$

and consequently, a non-zero irreducible representation π of $\Delta(K)$ and a non-zero vector $h \in H_\pi$ such that

$$\pi((\alpha e - x))(h) = 0.$$

If $\pi|A \neq 0$, then π is irreducible on A, and, since $K_{\pi(A)}[H] = H$, can be extended to an irreducible *-representation of $\Gamma(K)$ by 2.13 and 4.4. Now if $\alpha e - x$ has an inverse y in $\Gamma(K)$, it follows that

$$\pi(y)\pi(\alpha e - x) = I_{H_\pi}$$

which contradicts

$$\pi((\alpha e - x))(h) = 0.$$

If $\pi(A) = 0$, then $\pi(x) = 0$; thus, $\alpha\pi(e)(h) = 0$ which implies $\alpha h = 0$ and we again have a contradiction. Hence $\alpha \in \sigma_{\Gamma(K)}(x)$ and our proof is complete.

Up to now the functional calculus that we have developed has been for normal elements of $\Gamma(K)$. We will now develop a functional calculus for non-normal elements.

5.35. Let $x = (S,T)$ be a multiplier of K and Ω_0 an open subset of the complex plane. For each $a \in K$, the set

$$\sigma_{B(\mathcal{L}_a)}(\lambda_a(S)) \cup \sigma_{B(\mathcal{R}_{a*})}(\rho_a(T))$$

is compact and contained in $\sigma_{\Gamma(K)}((S,T))$. Consequently, we can find a bounded open set Ω with the following properties:

$$\sigma_{B(\mathcal{L}_a)}(\lambda_a(S)) \cup \sigma_{B(\mathcal{R}_{a*})}(\rho_a(T)) \subset \Omega \subset \Omega_0;$$

Ω has a finite number of components Ω_μ; each Ω_μ is bounded by a finite number of simple closed rectifiable curves $B_{\mu\nu}$; Ω has positive distance from the boundary of Ω_0. We assign positive orientation to each $B_{\mu\nu}$ in the usual manner and let $B = \cup B_{\mu\nu}$. We call B an oriented envelope of

$$\sigma_{B(\mathcal{L}_a)}(\lambda_a(S)) \cup \sigma_{B(\mathcal{R}_{a*})}(\rho_a(T))$$

with respect to Ω_0. Now let f be a complex-valued analytic function in Ω_0. Define

$$\tilde{f}(\lambda_a(S)) = \frac{1}{2\pi i} \int_B f(\nu)(\nu\lambda_a(I) - \lambda_a(S))^{-1} d\nu$$

and

$$\tilde{f}(\rho_a(T) = \frac{1}{2\pi i} \int_B f(\nu)(\nu\rho_a(I) - \rho_a(T))^{-1} d\nu,$$

where the integrals are in the usual Riemann-Stieltjes sense. By virtue of Cauchy's theorem, $\tilde{f}(\lambda_a(S))$ and $\tilde{f}(\rho_a(T))$ depend upon f and not on the domain Ω_0. Now define the linear maps $\tilde{f}(S):K \to K$ and $\tilde{f}(T):K \to K$ by the formulas

$$\tilde{f}(S)(a) = \tilde{f}(\lambda_a(S))(a)$$

and

$$\tilde{f}(T)(a) = \tilde{f}(\rho_a(T))(a)$$

for each $a \in K$. Denote the pair $(f(S),f(T))$ by $f((S,T))$.

5.36. Theorem. The pair $(f(S),f(T))$ is a multiplier of K and

$$(\lambda_a f(S),\rho_a f(T)) = (f(\lambda_a(S)), f(\rho_a(T)))$$

for each $a \in K$; therefore,

$$\sigma_{\Gamma(K)}(f((S,T))) = f(\sigma_{\Gamma(K)}((S,T))).$$

Moreover, if (S,T) is normal, then $\tilde{\tilde{f}}((S,T)) = f((S,T))$.

Proof. Let $u,v \in K$. By 3.1 there is an $a \in K^+$ such that $u \in \mathcal{R}_a$ and $v \in \mathcal{L}_a$. It follows from 5.5

$$u\tilde{f}(S)(v) = u(\tfrac{1}{2\pi i} \textstyle\int_B f(\nu)(\nu\lambda_v(I) - \lambda_v(S))^{-1}d\nu)(v)$$

$$= u(\tfrac{1}{2\pi i} \textstyle\int_B f(\nu)(\nu\lambda_a(I) - \lambda_a(S))^{-1}d\nu)(v)$$

$$= \tfrac{1}{2\pi i} \textstyle\int_B f(\nu)u(\nu\lambda_a(I) - \lambda_a(S))^{-1}d\nu(v)$$

$$= \tfrac{1}{2\pi i} \textstyle\int_B f(\nu)(\nu\rho_a(I) - \rho_a(T))^{-1}(u)d\nu)v$$

$$= (\tfrac{1}{2\pi i} \textstyle\int_B f(\nu)(\nu(\rho_u(I)) - \rho_u(T))^{-1}d\nu)(u)v$$

$$= \tilde{f}(T)(u)v.$$

Hence, $(\tilde{f}(S),\tilde{f}(T))$ is a multiplier of K. It is clear from what we have done so far that

$$(\lambda_a\tilde{f}(S), \rho_a\tilde{f}(T)) = (\tilde{f}(\lambda_a(S)), \tilde{f}(\rho_a(T))).$$

The proof that

$$\sigma_{\Gamma(K)}\tilde{f}((S,T)) = \tilde{f}(\sigma_{\Gamma(K)}((S,T)))$$

is identical to 5.19. Finally, to complete the proof, we need to show that $\tilde{f}((S,T)) = f((S,T))$ when (S,T) is normal. But this follows immediately from 5.14, the equality

$$(\lambda_a\tilde{f}(S),\ \rho_a\tilde{f}(T)) = (\tilde{f}(\lambda_a(S)),\ \tilde{f}(\rho_a(T))),$$

and the usual functional calculus for C*-algebras.

Remark. Since $\tilde{f}(x) = f(x)$ for all normal elements x in $\Gamma(K)$, we will use $f(x)$ instead of $\tilde{f}(x)$ for the non-normal elements and still be consistent.

For each $x \in \Gamma(K)$, let $\mathfrak{F}(x)$ denote the family of all complex valued functions f which are analytic on some neighborhood of $\sigma_{\Gamma(K)}(x)$. The neighborhood need not be connected and can depend on $f \in \mathfrak{F}(x)$.

5.37. <u>Theorem</u>. If f and g are in $\mathfrak{F}(x)$, h is in $\mathfrak{F}(g(x))$, and α, β are complex numbers, then the following statements hold:

(i) $\alpha f(x) + \beta g(x) = (\alpha f + \beta g)(x)$;

(ii) $f(x)g(x) = f \cdot g(x)$;

(iii) if f has a power series expansion $f(\nu) = \Sigma_{K=0}^{\infty} \alpha_K \nu^k$ valid in $\sigma_{\Gamma(K)}(x)$, then $f(x) = \Sigma_{k=0}^{\infty} \alpha_k x^k$;

(iv) if $\overline{f}(\nu) = \overline{f(\overline{\nu})}$, then $f \in F(x^*)$ and $\overline{f}(x^*) = f(x)^*$;

(v) $F(x) = h(g(x))$, where $F(\nu) = h(g(\nu))$.

<u>Proof</u>. The proof is similar to 5.20.

5.38. <u>Proposition</u>. Let $\{f_\lambda\}$ be in $\mathfrak{F}(x)$ and suppose that all the functions f_λ are analytic in some fixed neighborhood V of $\sigma_{\Gamma(K)}(x)$. Then if f_λ converges uniformly on compact subsets of V to the function f, $f_\lambda(x)$ converges to f(x) in the κ-topology.

Proof. The proof is similar to 5.17.

The remainder of this section is devoted to a Dauns-Hofmann theorem for $\Gamma(K)$. It has appeared in [26] and we include it here for completeness.

Throughout the remainder of this section, $\hat{A}$ will denote the spectrum of A.

5.39. Lemma. Let $x \in K$. Then $\{\pi \in \hat{A}: \pi(x) \neq 0\}$ is contained in a compact subset of $\hat{A}$.

Proof. Since K is spanned by its positive elements and K^+ is a face, it is enough to prove the assertion for those $x \in K$, $0 \leq x \leq 1$, for which there is a $y \in A^+$ such that $x = xy$ (see 2.3). But when

$$\pi(x) \neq 0, \pi \in \hat{A},$$

if follows that

$$\|\pi(y)\| \geq 1.$$

Consequently,

$$\{\pi \in \hat{A} | \pi(x) \neq 0\} \subseteq \{\pi \in \hat{A} | \|\pi(y)\| \geq 1\}$$

and the conclusion follows from [13, 3.37, p. 64].

Let $a \in K$ and B_a the hereditary C*-subalgebra of A generated by a. By virtue of 2.7 and 5.39, $\hat{B}_a$ is contained in a compact subset of $\hat{A}$. Hence we can prove the following lemma.

5.40. Lemma. Let f be a continuous complex valued function on $\hat{A}$. Then for every $a \in K$ there is $x \in K$ such that

$$\pi(x) = f(\pi)\pi(a)$$

for all $\pi \in \hat{A}$.

Proof. By the above remark $\hat{B}_a$ is contained in a compact subset of $\hat{A}$. Thus $f_{|\hat{B}_a}$ is bounded and by the Dauns-Hofmann theorem [16, p. 379] there is $x \in B_a \subseteq K$ (see 2.6) such that

$$\pi(x) = f(\pi)\pi(a), \ \pi \in \hat{B}_a.$$

Clearly, the above equality also holds for all $\pi \in \hat{A}$.

5.41. Lemma. Let $x \in \Gamma(K)$ and let P(A) be the set of all pure states of A. Then $\phi \to \phi(x)$ is w*-continuous on P(A).

Proof. Remark that $\phi(x)$ is defined by 4.5. Let $\phi_0 \in P(A)$ and let $\sqrt{3}/16 > \epsilon > 0$. Choose $a \in K$, $0 \leq a \leq 1$, such that

$$1 > \phi_0(a) > 1-\epsilon^2.$$

Denote

$$V = \{\phi \in P(A): |\phi(a) - \phi_0(a)| < 3\epsilon^2, \ |\phi(ax) - \phi_0(ax)| < \epsilon\}.$$

For $\phi \in V$ let π_ϕ be the corresponding irreducible *-representation on H_ϕ and let $\xi_\phi \in H_\phi$ be such that

$$\phi(z) = (\pi_\phi(z)\xi_\phi, \xi_\phi)$$

for all $z \in A$. Let M be the norm of the left multiplication by x^*x in $\mathcal{L}_a$ (see Section 3). Then

$$(\pi_\phi(a)(\xi_\phi),\xi_\phi) = \phi(a) > 1 - 4\epsilon^2,$$

thus

$$\|\pi_\phi(a^{1/2})\xi_\phi\| > (1-4\epsilon^2)^{1/2}.$$

From [13, 2.8.2, p. 43], it follows that there is $T_\phi \in \pi_\phi(A)$ such that

$$T_\phi \pi_\phi(a^{1/2})\xi_\phi = \xi_\phi$$

and

$$\|T_\phi\| \leq 3(1-4\epsilon^2)^{-1/2}.$$

Choose $b_\phi \in A$ such that

$$\pi_\phi(b_\phi) = T_\phi$$

and

$$\|b_\phi\| \leq 3(1-4\epsilon^2)^{-1/2} + 1.$$

We have

$$\phi(x^*x) = (\pi_\phi(x^*x)\xi_\phi,\xi_\phi) = (\pi_\phi(x^*x)\pi_\phi(b_\phi)\pi_\phi(a^{1/2})\xi_\phi,\xi_\phi)$$

$$= \phi(x^*xb_\phi a^{1/2}) \leq \|x^*xb_\phi a^{1/2}\| \leq [3(1-4\epsilon^2)^{-1/2} + 1]M < 7M.$$

Hence

$$|\phi(x)-\phi_0(x)| \leq |\phi(x)-\phi(ax)| + |\phi(ax)-\phi_0(ax)| +$$

$$+ |\phi_0(ax)-\phi_0(x)| \leq |\phi(1-a)x)| + |\phi_0((1-a)x)| + \epsilon \leq$$

$$\leq (\phi(x^*x)\phi((1-a)^2))^{1/2} + (\phi_0(x^*x)\phi_0((1-a)^2))^{1/2} + \epsilon \leq$$

$$\leq (7M)^{1/2}((\phi(1-a))^{1/2} + (\phi_0(1-a))^{1/2}) + \epsilon \leq (7M)^{1/2}(3\epsilon) + \epsilon.$$

Remark. If $\{y_\alpha\} \subseteq K$ and $y_\alpha \to y$ in the κ-topology and $\phi \in P(A)$, then

$$\pi_\phi(y_\alpha) \to \pi_\phi(y)$$

in the strong operator topology, since for any $\eta \in H_\phi$ there is $\xi \in H_\phi$ and $u \in K$ with

$$\pi_\phi(u)\xi = \eta.$$

Now,

$$\pi_\phi(y_\alpha)\eta = \pi_\phi(y_\alpha u)\xi \to \pi_\phi(y)\eta.$$

If x is in the centre of Z of $\Gamma(K)$, then for every $\phi \in P(A)$, $\pi_\phi(x)$ commutes with $\pi_\phi(A)$, thus is a scalar: $\pi_\phi(x) = \phi(x)$, and if π_{ϕ_1}, π_{ϕ_2} are equivalent we have $\phi_1(x) = \phi_2(x)$. Thus any $x \in Z$ defines a continuous function f_x on A by

$$f_x(\pi_\phi) = \phi(x).$$

5.42. Theorem. The mapping $x \to f_x$ is a *-isomorphism of Z onto $C(\hat{A})$.

Proof. Clearly, $x \to f_x$ is a *-homorphism. If $f_{x_1} = f_{x_2}$ for $x_1, x_2 \in Z$, then $\pi(x_1) = \pi(x_2)$ for every $\pi \in \hat{A}$. Thus $\pi(x_1 a) = \pi(x_2 a)$ for every $a \in K$, $\pi \in \hat{A}$, so $x_1 a = x_2 a$ for each $a \in K$. Similarly, $ax_1 = ax_2$ for all $a \in K$, hence $x_1 = x_2$ and the mapping is one to one.

Now let $f \in C(\hat{A})$. For each $x \in K$ let $T(x)$ be the element of K given by 5.40. Thus $\pi(T(x)) = f(\pi)\pi(x)$ for every $\pi \in \hat{A}$. Let $x,y \in K$. Then,

$$\pi(xT(y)) = f(\pi)\pi(x)\pi(y) = \pi(T(x)y),\ \pi \in \hat{A}.$$

Thus $z = (T,T) \in \Gamma(K)$ and clearly $z \in Z$. If $\{e_\lambda\}$ is an approximate identity from K, then

$$\phi(e_\lambda z) = (\pi_\phi(e_\lambda z)\xi_\phi,\xi_\phi) = f(\pi_\phi)\phi(e_\lambda)$$

for every $\phi \in P(A)$, so $\phi(z) = f(\pi_\phi)$ and $f = f_z$.

5.43. <u>Corollary</u>. Let $f \in C(\hat{A})$. Then for any $x \in \Gamma(K)$ there is a $y \in \Gamma(K)$ such that

$$\pi(y) = f(\pi)\pi(x)$$

for all $\pi \in \hat{A}$.

5.44. <u>Proposition</u>. The mapping $x \to f_x$ given by 5.42 is bicontinuous when Z is considered with the κ-topology and $C(\hat{A})$ with the compact open topology.

<u>Proof</u>. Let $x_\xi \xrightarrow{\kappa} x$ in Z. Let $S \subset \hat{A}$ be compact. For any $\pi_0 \in S$ there is $a_0 \in K^+$ such that $\pi_0(a_0) \neq 0$. Thus

$$\{\pi \in \hat{A}:\|\pi(a_0)\| > \tfrac{1}{2}\|\pi_0(a_0)\|\}$$

is a neighborhood of π_0 (cf. [13, 3.3.2, p. 63]). Thus there are

$$\{a_i\}_{i=1}^n \subset K^+ \text{ and } \{\pi_i\}_{i=1}^n \subset S$$

such that

$$S \subset \bigcup_{i=1}^{n} \{\pi \in \hat{A} : \|\pi(a_i)\| > \tfrac{1}{2}\|\pi_i(a_i)\|\}.$$

Put

$$\lambda = \min(\tfrac{1}{2}\|\pi_i(a_i)\|) > 0 \text{ and } a = \Sigma_{i=1}^{n} a_i.$$

Then for any $\pi \in S$ we have

$$\|\pi(a)\| > \tfrac{1}{2}\|\pi_i(a_i)\| \geq \lambda$$

(for a suitable i). Now,

$$\|x_\alpha a - xa\| \to 0,$$

so

$$\sup_{\pi \in \hat{A}} |\pi(x_\alpha)\pi(a) - \pi(x)\pi(a)| \to 0.$$

Thus for any $\epsilon > 0$ there is α_0 such that $\alpha > \alpha_0$ implies

$$\begin{aligned} |f_{x_\alpha}(\pi) - f_x(\pi)| &= \|(f_{x_\alpha}(\pi) - f_x(\pi))\pi(a)\| / \|\pi(a)\| \\ &= \|\pi(x_\alpha)\pi(a) - \pi(x)\pi(a)\| / \|\pi(a)\| \\ &\leq \epsilon/\lambda. \end{aligned}$$

Assume now that $f_{x_\alpha} \to f_x$ in the compact open topology of $C(\hat{A})$. Let $a \in K$. By 5.39, the set $\{\pi \in \hat{A} : \pi(a) \neq 0\}$ is contained in a compact subset of $\hat{A}$. Thus on it f_{x_α} converges uniformly to f_x. Let $\epsilon > 0$; there is α_0 such that $\alpha > \alpha_0$ implies

$$|f_{x_\alpha}(\pi) - f_x(\pi)| < \epsilon,\ \pi \in \hat{A},\ \pi(a) \neq 0.$$

Hence

$$\|x_\alpha a - xa\| = \sup_{\pi \in \hat{A}} \|\pi(x_\alpha)\pi(a) - \pi(x)\pi(a)\| =$$

$$= \sup_{\pi(a) \neq 0} \|\pi(a)\| \, |f_{x_\alpha}(\pi) - f_x(\pi)| \leq \|a\| \epsilon .$$

CHAPTER 6. THE DUAL OF $\Gamma(K)$

In this chapter we shall give a representation for $\Gamma(K)'$, the dual of $\Gamma(K)$, the latter being considered with its κ-topology. As in previous chapters, A will be a C*-algebra and K its Pedersen ideal. If $f \in A'$, the Banach space dual of A, and $a \in A$ then $a \cdot f$ and $f \cdot a$ denote the linear functionals $x \to f(xa)$ and $x \to f(ax)$, respectively.

6.1. Theorem. $\Gamma(K)' = \{a \cdot f + g \cdot a : a \in K^+,\ f,g \in A',$ $\|f\| \leq 1,\ \|g\| \leq 1\}$.

Proof. For $a \in K^+$ denote

$$U_a = \{x \in \Gamma(K) : \|xa\| \leq 1\},$$

$$V_a = \{x \in \Gamma(K) : \|ax\| \leq 1\},$$

$$W_a = U_a \cap V_a$$

and

$$\Sigma = \{f \in A' : \|f\| \leq 1\}.$$

Since $\{W_a : a \in K^+\}$ is a neighborhood base at 0,

$$\Gamma(K)' = \bigcup \{W_a^0 : a \in K^+\},$$

where W_a^0 is the polar of W_a taken in the algebraic dual of $\Gamma(K)$. Suppose for the moment that the following facts have already been established:

(i) U_a and V_a are absolutely convex sets closed in the weak topology of $\Gamma(K)$ induced by $\Gamma(K)'$;

(ii) $U_a^0 \subseteq \{a\cdot f : f \in \Sigma\}$ and $V_a^0 \subseteq \{f\cdot a : f \in \Sigma\}$;

(iii) The set $a\cdot\Sigma + \Sigma\cdot a = \{a\cdot f + g\cdot a : f,g \in \Sigma\}$ is a $\sigma(\Gamma(K)', \Gamma(K))$ closed absolutely convex subset of $\Gamma(K)'$.

Since W_a is weakly closed, W_a^0 is the smallest $\sigma(\Gamma(K)', \Gamma(K))$ closed absolutely convex subset of $\Gamma(K)'$ that contains $U_a^0 \cup V_a^0$. Hence, by (ii) and (iii),

$$W_a^0 \subseteq a\cdot\Sigma + \Sigma\cdot a$$

and the proof is complete provided (i), (ii) and (iii) are true.

Now we shall prove (i). Clearly, it will suffice to show that U_a is weakly closed because a similar argument can be applied to V_a. Let y be a weak limit point of U_a. Then there exists a net $\{x_\alpha\} \subset U_a$ such that $x_\alpha \to y$ weakly. It follows that

$$\|ya\| = \sup\{|f(ya)| : f \in \Sigma\} = \sup\{|(a\cdot f)(y)| : f \in \Sigma\} =$$

$$= \sup\{\lim_\alpha |(a\cdot f)(x_\alpha)| : f \in \Sigma\} =$$

$$= \sup\{\lim_\alpha |f(x_\alpha a)| : f \in \Sigma\} \leq 1,$$

since $a\cdot f \in \Gamma(K)'$ and $\{x_\alpha\} \subset U_a$. Hence $y \in U_a$ and (i) is established.

Next we shall prove (ii). Let $f \in U_a^0$; then $|f(x)| \leq 1$ for each $x \in U_a$ or, equivalently,

$$|f(x)| \leq \|xa\|$$

for each $x \in \Gamma(K)$. Now, on the normed linear subspace

$$\{xa : x \in \Gamma(K)\}$$

of A define the linear functional g by $g(xa) = f(x)$. Clearly, g is well defined and bounded by 1. Hence g has a norm preserving extension g to all of A. It follows that $a \cdot g = f$, since K is κ-dense in $\Gamma(K)$ and

$$|f(x) - f(x_\alpha)| \leq \|(x - x_\alpha)a\|$$

for $x \in \Gamma(K)$ and $x_\alpha \in K$, and this implies

$$U_a^0 \subset \{a \circ f : f \in \Sigma\}$$

Similarly,

$$V_a^0 \subset \{f \cdot a : f \in \Sigma\}.$$

Finally, we shall show that (iii) holds. It is clear that $a \cdot \Sigma + \Sigma \cdot a$ is absolutely convex, so let h be a $\sigma(\Gamma(K)', \Gamma(K))$ limit point of $a \cdot \Sigma + \Sigma \cdot a$. This means that there exists a net $\{a \cdot f_\alpha + g_\alpha \cdot a\}$, with $f_\alpha, g_\alpha \in \Sigma$, such that

$$h(x) = \lim_\alpha [f_\alpha(xa) + g_\alpha(ax)]$$

for each $x \in \Gamma(K)$. Since Σ is compact in the $\sigma(A',A)$ topology, there exist $f, g \in \Sigma$ and subnets of $\{f_\alpha\}$ and $\{g_\alpha\}$ which converge $\sigma(A',A)$ to f and g respectively. To simplify the notation we shall assume that $\{f_\alpha\}$ and $\{g_\alpha\}$ are $\sigma(A',A)$ convergent. It follows that

$$h(x) = \lim [f_\alpha(xa) + g_\alpha(ax)] =$$

$$= \lim_\alpha f_\alpha(xa) + \lim_\alpha g_\alpha(ax) = f(xa) + g(ax) = (a \cdot f + g \cdot a)(x)$$

for every $x \in \Gamma(K)$ and this completes the proof.

6.2. <u>Corollary</u>. $\Gamma(K)'$ is uniformly dense in A'.

6.3. A positive linear functional on $\Gamma(K)$ is not necessarily κ-continuous. Indeed let X be a pseudo-compact, non-compact, locally compact Hausdorff space, $A = C_0(X)$ and $p \in \beta X \setminus X$. The evaluation at p is a non-trivial positive functional on $\Gamma(K) = C(X) = C_b(X)$, but it is not κ-continuous since it vanishes on $K = C_{00}(X)$. For those positive functionals on $\Gamma(K)$ which are κ-continuous we shall obtain a special representation.

6.4. Lemma. Let f be a positive κ-continuous functional on $\Gamma(K)$, π the associate *-representation of A, ξ the corresponding cyclic vector. Then there exists a C*-subalgebra $B \subset K$ such that $\xi \in \overline{\pi(B)\xi}$.

Proof. Let $b \in K^+$ satisfy:

$$\|bx\|, \|xb\| \leq 1 \Rightarrow |f(x)| \leq 1;$$

and denote by B the C*-subalgebra generated by b. Then $B \subseteq K$ by 2.6. Let $\{e_\lambda\}$ be a positive approximate identity for B satisfying $\|e_\lambda\| \leq 1$ for each λ. Given $\epsilon > 0$, there exists λ_0 such that if $\lambda > \lambda_0$, then

$$\|b(1 - e_\lambda)(1 - e_\lambda)\| \leq \epsilon^2.$$

Thus,

$$0 \leq f((1 - e_\lambda)(1 - e_\lambda)) < \epsilon^2$$

and

$$\|\xi - \pi(e_\lambda)\xi\| \leq \epsilon$$

if $\lambda > \lambda_0$.

6.5. Proposition. Let f be a positive κ-continuous functional on $\Gamma(K)$. Then there is a positive functional $g \in A'$ and $a \in K$ such that $f(x) = g(a^*xa)$ for all $x \in \Gamma(K)$.

Proof. Let π, ξ and B be as in 6.4. Denote

$$H_1 = \overline{\pi(B)\xi} \subset H_\pi.$$

Then $z \to \pi(z)_{|H_1}$ is a cyclic representation of B into $B(H_1)$ and $\xi \in H_1$. By the Cohen-Hewitt factorization theorem [18, Theorem 2.5, p. 151], there is $a \in B \subset K$ and $\eta \in H_1$ such that $\pi(a)\eta = \xi$. Denoting by g the functional

$$g(x) = \langle \pi(x)\eta, \eta \rangle$$

we get

$$f(x) = \langle \pi(x)\xi, \xi \rangle = \langle \pi(a^*xa)\eta, \eta \rangle = g(a^*xa)$$

for each $x \in A$. Since f and $x \to g(z^*xa)$ are κ-continuous, we obtain $f(x) = g(a^*xa)$ for every $x \in \Gamma(K)$.

Question. Is $\Gamma(K)'$ the linear span of its positive cone?

CHAPTER 7. HOMOMORPHISMS OF $\Gamma(K)$

It has been remarked in 2.13 that if ϕ is a *-homomorphism of a C*-algebra A onto a C*-algebra B, then ϕ can be extended to a *-homomorphism $\overline{\phi}$ of $\Gamma(K_A)$ in $\Gamma(K_B)$ given by

$$\overline{\phi}(x)\phi(a) = \phi(xa),$$

$$\phi(a)\overline{\phi}(x) = \phi(ax)$$

for $x \in \Gamma(K_A)$, $a \in K_A$. We shall establish in this chapter conditions under which the extension $\overline{\phi}$ is onto $\Gamma(K_B)$. In view of 4.1, this problem can be regarded as a non-commutative extension of Tietze's theorem. For bounded multipliers the problem was considered in [3, Theorem 4.2]. It is shown there that if A is separable then $\overline{\phi}$ maps M(A) onto M(B). The last section of this chapter deals with *-isomorphisms between $\Gamma(K_A)$ and $\Gamma(K_B)$.

Throughout this chapter, unless otherwise stated, A and B will denote C*-algebras, ϕ a *-homomorphism of A onto B and $\overline{\phi}$ its extension to $\Gamma(K_A)$ defined in 2.13.

7.1. <u>Proposition</u>. If $\overline{\phi}$ maps $\Gamma(K_A)$ onto $\Gamma(K_B)$, then it maps M(A) onto M(B).

Proof. Suppose $y \in M(B)^+$. By assumption there is $x \in \Gamma(K_A)^+$ such that $\overline{\phi}(x) = y$. Let f be the real valued function defined by

$$f(t) = \begin{cases} 0 & t \leq 0, \\ t & 0 < t \leq \|y\|, \\ \|y\| \;, & \|y\| \leq t. \end{cases}$$

It is a consequence of 5.19 and 5.30 that $f(x) \in M(A)$. Consider now a sequence of polynomials $\{p_n\}_{n=1}^{\infty}$ which converges to f in the compact open topology. By 5.17, $\{p_n(x)\}_{n=1}^{\infty}$ converges to $f(x)$ in the κ-topology and by 3.11 we have

$$\overline{\Phi}(f(x)) = \kappa\text{-}\lim_{n\to\infty} \overline{\Phi}(p_n(x)) = \lim_{n\to\infty} p_n(\overline{\Phi}(x)) =$$

$$= \lim_{n\to\infty} p_n(y) = y.$$

This concludes the proof of the proposition.

7.2. <u>Lemma</u>. Let $x \in \Gamma(K_A)$ and F be a compact subset of $\hat{A}$. Then

$$\sup \{\|\pi(x)\| : \pi \in F\} < \infty .$$

<u>Proof</u>. We may suppose $x \in \Gamma(K_A)^+$. For each $\pi \in F$ choose $a_\pi \in K$ such that

$$\|a_\pi\| \leq 1 \text{ and } \|\pi(a_\pi)(1_\pi)\| > 4^{-1}$$

(cf. [13, 2.8.3, p. 44]). The state f_π given by

$$f_\pi(z) = \langle \pi(z)(1_\pi), 1_\pi \rangle$$

belongs to $P(A)$, the set of all pure states of A. Put

$$U_\pi = \{f \in P(A) : |f_\pi(a_\pi^* a_\pi) - f(a_\pi^* a_\pi)| < 8^{-1}\}$$

and let V_π be the image of U_π by the canonical map of $P(A)$ onto $\hat{A}$. V_π is a neighborhood of π [13, 3.4.11, p. 68]. Let $\{V_{\pi_j}\}_{j=1}^{n}$ be a sub-cover of the cover $\{V_\pi : \pi \in F\}$ of F and denote $a_j = a_{\pi_j}$. The multiplication on the left by x is bounded on the closed left ideal $\mathcal{L}$ of A generated by $\{a_j\}_{j=1}^{n}$. Indeed, one easily sees that $\mathcal{L}$ is the closed left ideal of A generated by some $a \in K$ and the assertion follows from 3.4. Suppose α satisfies

$$\|xy\| \leq \alpha \|y\|$$

for each $y \in \mathfrak{L}$.

Now let $\pi \in F$, which implies for some π_j, $\pi \in V\pi_j$. There is $f \in U\pi_j$ such that $\pi = \pi_f$. We have $f(a^*_j a_j) > 8^{-1}$, hence

$$\|\pi(a_j)1_\pi\| > 2^{-3/2}.$$

Let $h \in H_\pi$ with $\|h\| = 1$. By [13, 2.8.2, p. 43] there is $b_h \in A$ with $\|b_h\| \leq (24)(2^{1/2})$ such that

$$\pi(b_h)\pi(a_j)(1\pi) = h.$$

Thus

$$\langle \pi(x)h,h \rangle = \langle \pi(a_j^*)\pi(b_h^*)\pi(x)\pi(b_h)\pi(a_j)(1_\pi),1_\pi \rangle$$

$$\leq |a_j^* \, b_h^* \, x \, b_h \, a_j| \leq 2(24)^2\alpha.$$

We have, therefore,

$$\|\pi(x)\| \leq 2(24)^2\alpha$$

for each $\pi \in F$ and the proof is complete.

7.3. Theorem. Suppose A has a countable approximate identity and Hausdorff spectrum. Then

$$\overline{\Phi}(\Gamma(K_A)) = \Gamma(K_B).$$

Proof. Let us denote $J = \phi^{-1}(0)$. We shall identify B with A/J and its spectrum with

$$\hat{A}_J = \{\pi \in \hat{A}: \pi(J) = 0\}.$$

By virtue of [37, Proposition 3.1] there is a countable increasing positive approximate identity $\{e_n\}_{n=1}^{\infty}$ for A such that

$$e_{n+1}e_n = e_n \text{ and } \|e_n\| = 1$$

for each n. It follows from 2.3 that $\{e_n\}_{n=1}^{\infty} \subset K_A$. Let E_n denote the closure of

$$\{\pi \in \hat{A}: \|\pi(e_n)\| > 0\}.$$

By 5.39, E_n is compact. Moreover, $E_n \subset E_{n+1}$ and, since $\{e_n\}_{n=1}^{\infty}$ is an approximate identity for A, it follows from [13, 3.3.2, p. 63] that every compact subset of $\hat{A}$ is contained in some E_n.

Now let $t \in \Gamma(K_B)^+$. Since $\hat{A}_J$ is a closed subset of $\hat{A}$, the sets $E_n \cap \hat{A}_J$ are compact and therefore, by 7.2, the numbers

$$r_n = \sup\{\|\pi(t)\| : \pi \in E_n \cap \hat{A}_J\}$$

are finite. The sequence $\{\phi(e_n)\}_{n=1}^{\infty}$ is an approximate identity for A/J, therefore,

$$\{t^{1/2}\phi(e_n)t^{1/2}\}_{n=1}^{\infty}$$

converges to t in the κ_B-topology. It follows that there is a subsequence $\{e_{n_k}\}_{k=1}^{\infty}$ for which the following inequality holds:

$$\text{(i)} \quad \|t^{1/2}\phi(e_{n_{k+1}} - e_{n_k})t^{1/2}\phi(e_k)\| < 2^{-k},$$

for $k = 1,2,\ldots$.

We shall now construct an increasing sequence $\{x_k\}_{k=1}^{\infty}$ in K_A^+ that satisfies the following:

(1) $\phi(x_k) = t^{1/2}\phi(e_{n_k})t^{1/2}$;

(2) $\{\pi \in \hat{A}:\|\pi(x_k)\| > 0\} \subset E_{n_k}$;

(3) $\sup\{\|\pi(x_k)\|:\pi \in E_{n_j}\} < r_{n_j} + 1$ for $k,j = 1,2,\ldots$;

(4) $\|(x_{k+1} - x_k)e_k\| < 2^{-k}$.

Note that

$$\|t^{1/2}\phi(e_{n_1}^{1/4})\| \leq r_{n_1}^{1/2} ,$$

so choose $x \in K_A$ for which

$$\pi(x) = \pi(e_{n_1}^{1/4})t^{1/2}$$

and

$$\|x\| \leq (r_{n_1} + 1)^{1/2}.$$

Set

$$x_1 = x^*e_{n_1}^{1/2}x.$$

Clearly, x_1 satisfies the appropriate conditions from above. Suppose now $x_1,x_2,\ldots,x_k$ are elements in K_A^+, with $x_1 \leq x_2 \leq \cdots \leq x_k$, that satisfy the conditions (1)-(4). Choose w in K_A^+ so that

$$\phi(w) = t^{1/2}\phi(e_{n_{k+1}} - e_{n_k})t^{1/2}$$

and, as before, we may suppose also that

$$\{\pi \in \hat{A} : \|\pi(w)\| > 0\} \subset E_{n_{k+1}} .$$

Now define the sets

$$F_j = \{\pi \in E_{n_j} : \|\pi(x_k + w)\| \geq r_{n_j} + 1\}$$

for $1 \leq j \leq k + 1$ and

$$F_0 = \{\pi \in \hat{A} : \|\pi(we_k)\| \geq 2^{-k}\}.$$

We now wish to show that $F = \bigcup_{j=0}^{k+1} F_j$ is a compact subset of $\hat{A}$ that does not meet $\hat{A}_J$. The fact that F is compact follows from [13, 3.3.7, p. 64]. Let $\pi \in \hat{A}_J$. Since

$$\|\pi(we_k)\| = \|\pi(t^{1/2})\pi(e_{n_{k+1}} - e_{n_k})\pi(t^{1/2})\pi(e_k)\| \leq$$

$$\leq \|t^{1/2}\phi(e_{n_{k+1}} - e_{n_k})t^{1/2}\phi(e_k)\| < 2^{-k}$$

by virtue of (i), it follows that $\pi \notin F_0$. If $\pi \in E_{n_j}$, then

$$\|\pi(x_k + w)\| = \|\pi(t^{1/2})\pi(e_{n_{k+1}})\pi(t^{1/2})\| \leq \|\pi(t)\| \leq r_{n_j}.$$

Hence, $\pi \notin F_j$. Thus F does not meet $\hat{A}_J$.

Since $\hat{A}$ is locally compact Hausdorff, there is a continuous function f on $\hat{A}$ with values in [0,1] such that $f(\hat{A}_J) = \{1\}$ and $f(F) = \{0\}$. By the Dauns-Hofmann theorem [11, III, 3.5, 3.9 and III, § 5], [16, § 7] there is an element $f \cdot w$ in A with the property:

$$\pi(f \cdot w) = f(\pi)\pi(w)$$

for each $\pi \in \hat{A}$. Let

$$x_{k+1} = x_k + f \cdot w.$$

Clearly x_{k+1} is in K_A^+ and

$$x_{k+1} \geq x_k.$$

Moreover, it is clear that the conditions (1) and (2) hold. Now let $\pi \in E_{n_j}$. By virtue of (2) we may assume $j \leq k + 1$. If $\pi \in F_j$, then

$$\|\pi(x_{k+1})\| = \|\pi(x_k) + f(\pi)\pi(w)\| = \|\pi(x_k)\| < r_{n_j} + 1.$$

If $\pi \notin F_j$, then

$$\|\pi(x_{k+1})\| = \|\pi(x_k) + f(\pi)\pi(w)\| \leq \|\pi(x_k+w)\| < r_{n_j} + 1.$$

By [13, 3.3.9, p. 65], the map

$$\pi \to \|\pi(x_{k+1})\|$$

attains

$$\sup\{\|\pi(x_{k+1})\| : \pi \in E_{n_j}\}$$

on E_{n_j}. Hence condition (3) holds. Since condition (4) can be verified in a similar manner, we have completed our induction.

We claim that $\{x_k\}_{k=1}^{\infty}$ is κ-Cauchy. Let $a \in K_A^+$ and $\epsilon > 0$. Recall that by 5.39, there is a natural number p for which

$$\{\pi \in A : \|\pi(a)\| > 0\} \subset E_p.$$

Choose an integer $N > p$ so that

$$\|a - e_N a\| < \epsilon/4(r_{n_p} + 1)$$

and

$$\Sigma_{s=N}^{\infty} 2^{-s} < \epsilon/2\|a\|.$$

It follows that for $r > s \geq N$,

$$\|(x_r - x_s)a\| \leq \|(x_r - x_s)e_N a\| + \|(x_r - x_s)(a - e_N a)\| \leq$$

$$\leq \Sigma_{k=s}^{r-1} \|(x_{k+1} - x_k)e_N\|\|a\| + \sup\{\|\pi(x_r - x_s)\pi(a - e_N a)\| : \pi \in E_p\} \leq$$

$$\leq \Sigma_{k=s}^{r-1} \|(x_{k+1} - x_k)e_k\|\|a\| + 2(r_{n_p} + 1)\|a - e_N a\| \leq \Sigma_{k=s}^{\infty} 2^{-k}\|a\| + \epsilon/2 < \epsilon.$$

Hence $\{x_k\}_{k=1}^{\infty}$ is κ_A-Cauchy. Since $\Gamma(K_A)$ is complete there is $x \in \Gamma(K_A)$ such that $x = \kappa\text{-}\lim_{k\to\infty} x_k$. Clearly $\phi(x) = t$. Due to 5.25, the proof is complete.

In contrast to the case of bounded multipliers investigated in [3, Theorem 4.2], the mere separability of A does not ensure that

$$\overline{\phi}(\Gamma(K_A)) = \Gamma(K_B).$$

7.4. <u>Example</u>. Let $\{H_n\}_{n=1}^{\infty}$ be a sequence of separable, infinite dimensional Hilbert spaces and $H = \Sigma \oplus H_n$, the Hilbert sum. For each n, let P_n be the orthogonal projection of H onto H_n. Then let C be the commutative C*-algebra generated by $\{P_n\}_{n=1}^{\infty}$. Note that every element $x \in C$ is of the form

$$x = \Sigma_{n=1}^{\infty} \lambda_n P_n$$

with

$$\{\lambda_n\}_{n=1}^{\infty} \in c_0 = C_0(N),$$

N being the set of natural numbers. Define $A = C + B_0(H)$. Since $B_0(H)$

is a uniformly closed two-sided ideal of B(H), A is a uniformly closed subalgebra of B(H) [13, 1.8.4, p. 18]. Thus A is a separable C*-algebra. Since $C \cap B_0(H) = 0$, the map $\phi: A \to C_0(N)$ given by

$$\phi(\Sigma_{n=1}^{\infty} \lambda_n P_n + R) = \{\lambda_n\}_{n=1}^{\infty}$$

is a well defined *-homomorphism of A onto $B = C_0(N)$. Note that $\Gamma(K_B) = C(N)$, the algebra of all sequences of complex numbers. On the other hand, A is a PCS-algebra, that is $\Gamma(K_A) = M(A)$. This is so since

$$K_A[H] = B_0(H)[H] = H$$

(cf. 4.4). Therefore,

$$\overline{\phi}(\Gamma(K_A)) = M(B) = C_b(N) = \ell_\infty \neq C(N).$$

It is easily seen that A is a type I C*-algebra.

7.5. A positive approximate identity $\{e_\lambda : \lambda \in \Lambda\}$ for the C*-algebra A is called well behaved if:

(1) $\|e_\lambda\| \leq 1$ for each $\lambda \in \Lambda$;

(2) $e_{\lambda_2} e_{\lambda_1} = e_{\lambda_1}$ if $\lambda_2 > \lambda_1$, $\lambda_2 \neq \lambda_1$;

(3) for every strictly increasing sequence $\{\lambda_n\}_{n=1}^{\infty}$ in Λ and every $\lambda \in \Lambda$ there is a positive integer N such that $e_\lambda(e_{\lambda_n} - e_{\lambda_m}) = 0$ whenever $n, m \geq N$ (cf. [37]).

7.6. A subset F of the spectrum $\hat{A}$ is called a Hausdorff set if F is closed and for each $\pi \in \hat{A} \setminus F$ there exists a real-valued continuous function f_π on $\hat{A}$ such that $f_\pi(\pi) \neq 0$ and $f_\pi(F) = \{0\}$.

For future reference we mention the following lemma, the proof of which is routine.

7.7. <u>Lemma</u>. Suppose F_1 is a Hausdorff set in $\hat{A}$ and F_2 is a compact subset of $\hat{A} \setminus F_1$. Then there exists a continuous function $f:\hat{A} \to [0,1]$ such that $f(F_1) = \{0\}$ and $f(F_2) = \{1\}$.

7.8. <u>Theorem</u>. Let $J = \phi^{-1}(0)$. Suppose A has a well behaved approximate identity, B has a countable approximate identity and $\hat{A}_J$ is a Hausdorff set. Then $\overline{\Phi}(M(A)) = M(B)$.

<u>Proof</u>. As in 7.3, we shall make the following identifications: $B = A/J$, $\hat{B} = \hat{A}_J$. Let $y \in M(A/J)^+$ and suppose $\{e_\lambda : \lambda \in \Lambda\}$ is a well behaved approximate identity for A. Since $\{\phi(e_\lambda) : \lambda \in \Lambda\}$ is an approximate identity for A/J, there exists a strictly increasing sequence $\{\lambda_n\}_{n=1}^{\infty} \subset \Lambda$ such that $\{\phi(e_{\lambda_n})\}$ is an approximate identity for A/J. Moreover, by dropping to a subsequence if necessary, we may assume that

$$(*) \qquad \| [\phi(e^2_{\lambda_{n+1}}) y\, \phi(e_{\lambda_{n+1}}) - \phi(e^2_{\lambda_n}) y\, \phi(e_{\lambda_n})]\, \phi(e_{\lambda_{n-1}}) \| < 2^{-n},$$

for $n = 2,3,\ldots$.

We shall now define a sequence $\{x_n\}_{n=1}^{\infty}$ in K_A^+ for which the following hold:

(1) $\|x_n\| < 2\|y\|$;

(2) $x_n = x_n e_{\lambda_{n+2}}$;

(3) $\phi(x_n) = \phi(e^2_{\lambda_{n+1}}) y \phi(e^2_{\lambda_{n+1}})$;

(4) $\|(x_{n+1} - x_n) e_{\lambda_n}\| < 2^{-n-1}$, $n = 1,2,\ldots$.

Since

$$\phi(K_A^+) = K_{A/J}^+,$$

there exists an element $z_1 \in K_A^+$ such that

$$\|z_1\| \leq 2\|y\| \text{ and } \phi(z_1) = \phi(e_{\lambda_2})y\,\phi(e_{\lambda_2}).$$

Now set

$$x_1 = e_{\lambda_2}\, z\, e_{\lambda_2}.$$

Clearly, x_1 satisfies (1), (2) and (3). Suppose $x_1, x_2, \ldots, x_n$ have already been defined satisfying (1)-(4). As before, there exists an element z_{n+1} in K_A^+ such that

$$\|z_{n+1}\| < 2\|y\|$$

and

$$\phi(z_{n+1}) = \phi(e_{\lambda_{n+2}})y\,\phi(e_{\lambda_{n+2}}).$$

Set

$$x'_{n+1} = e_{\lambda_{n+2}}\, z_{n+1}\, e_{\lambda_{n+2}}.$$

It is clear that x'_{n+1} satisfies (1)-(3). Now let

$$F = \{\pi \in \hat{A} : \|\pi(x'_{n+1} - x_n)\pi(e_{\lambda_n})\| \geq 2^{-n-1}\}$$

which is a compact subset of $\hat{A}$. Moreover, by (*), F and $\hat{A}_J$ are disjoint. Consequently, by 7.7, there exists a continuous function $f : \hat{A} \to [0,1]$ such that $f(F) = \{0\}$ and $f(\hat{A}_J) = \{1\}$. By the Dauns-Hofmann theorem [16, § 7], there are elements $f \cdot x_n$ and $f \cdot x'_{n+1}$ in K_A^+ such that

$$\pi(f \cdot x_n) = f(\pi)\pi(x_n)$$

and

$$\pi(f \cdot x'_{n+1}) = f(\pi)\pi(x'_{n+1})$$

for each $\pi' \in \hat{A}$. Now set

$$x_{n+1} = f \cdot x'_{n+1} - f \cdot x_n + x_n.$$

It is easy to establish that x_{n+1} satisfies (1)-(4). Hence the induction is complete.

We now wish to show that $\{x_n\}_{n=1}^{\infty}$ is κ_A-Cauchy. Observe that $\{x_n a\}_{n=1}^{\infty}$ is uniformly Cauchy for each $a \in K_A$ if and only if $\{x_n e_\lambda\}_{n=1}^{\infty}$ is uniformly Cauchy for each $\lambda \in \Lambda$. Let $\mu \in \Lambda$. Since $\{e_\lambda : \lambda \in \Lambda\}$ is a well behaved approximate identity, there exists an integer N such that for $m,n \geq N$ we have

$$e_{\lambda_m} e_\mu = e_{\lambda_n} e_\mu$$

So, for $m,n \geq N$, we get from (2) and (4),

$$\|(x_n - x_m)e_\mu\| = \|x_n e_{\lambda_{n+2}} e_\mu - x_m e_{\lambda_{m+2}} e_\mu\| =$$

$$= \|x_n e_{\lambda_N} e_\mu - x_m e_{\lambda_N} e_\mu\| \leq \|(x_n - x_m)e_{\lambda_N}\| \leq$$

$$\leq \Sigma_{k=m}^{n-1} \|(x_{k+1} - x_k)e_{\lambda_N}\| = \Sigma_{k=m}^{n-1} \|(x_{k+1} - x_k)e_{\lambda_k}e_{\lambda_N}\| \leq$$

$$\leq \Sigma_{k=m}^{n-1} \|(x_{k+1} - x_k)e_{\lambda_k}\| \leq \Sigma_{k=m}^{n-1} 2^{-k-1} .$$

Thus, $\{x_n\}_{n=1}^{\infty}$ is κ_A-Cauchy and must converge to some element $x \in M(A)^+$ by virtue of (1). Since $\overline{\phi}$ is κ-continuous, it follows from (3) that $\overline{\phi}(x) = y$ and the proof is complete.

7.9. If $\Gamma(K_A)$ and $\Gamma(K_B)$ are *-isomorphic, it does not necessarily follow that A and B are *-isomorphic. Indeed, let Ω be the first uncountable ordinal, $A = C_0([0,\Omega))$, $B = C([0,\Omega])$. The identity map is a *-isomorphism of $\Gamma(K_A) = C([0,\Omega])$ onto $\Gamma(K_B) = C([0,\Omega])$. Obviously, A and B are not *-isomorphic.

7.10. Theorem. Let A and B be C*-algebras and ψ a κ-bicontinuous *-isomorphism of $\Gamma(K_A)$ onto $\Gamma(K_B)$. Then $\psi(A) = B$; in particular A and B are *-isomorphic.

Proof. If $x \in M(A)$ is normal, then its spectrum with respect to $\Gamma(K_A)$ is bounded. Thus $\psi(x)$ is normal and its spectrum with respect to $\Gamma(K_B)$ is bounded. It follows from 5.30 that $\psi(x) \in M(B)$. Hence $\psi(M(A)) \subseteq M(B)$. The converse inclusion can be proved similarly, so $\psi(M(A)) = M(B)$.

Let now $a \in K_A^+$. For any $y \in \Gamma(K_B)$, $a\psi^{-1}(y)$ and $\psi^{-1}(y)a$ are in K_A, so $\psi(a)y$ and $y\psi(a)$ are in $M(B)$. Moreover, the set

$$\{y \in \Gamma(K_B) : \|\psi(a)y\| \leq 1, \|y\psi(a)\| \leq 1\} =$$

$$= \{y \in \Gamma(K_B) : \|a\psi^{-1}(y)\| \leq 1, \|\psi^{-1}(y)a\| \leq 1\}$$

being the image by ψ of

$$\{x \in \Gamma(K_A) : \|ax\| \leq 1, \|xa\| \leq 1\},$$

is a neighborhood of the origin in the κ_B-topology. Hence, there exists an element $b \in K_B^+$ such that

$$\{y \in \Gamma(K_B) : \|by\| \leq 1, \|yb\| \leq 1\} \subseteq \{y \in \Gamma(K_B) : \|\psi(a)y\| \leq 1, \|y\psi(a)\| \leq 1\}.$$

If a selfadjoint $y \in \Gamma(K_B)$ satisfies

$$\|yb^2y\| \leq 1,$$

then

$$\|\psi(a)y\| \leq 1 \text{ and } \|y\psi(a)\| \leq 1$$

by the previous inclusion, so

$$\|y\psi(a)^2y\| \leq 1.$$

Hence,

$$\|y\psi(a)^2y\| \leq \|yb^2y\|$$

for any selfadjoint $y \in \Gamma(K_B)$.

Let $\{u_\lambda\}$ be a positive right approximate identity for the norm closed left ideal $\mathcal{L}$ generated by b in B (cf. [13, 1.7.3, p. 16]). We have

$$\|\psi(a) - \psi(a)u_\lambda\|^2 = \|(1 - u_\lambda)\psi(a)^2(1 - u_\lambda)\| \leq \|(1 - u_\lambda)b^2(1 - u_\lambda) =$$

$$= \|b - bu_\lambda\|^2.$$

Thus $\psi(a) = \lim_\lambda \psi(a)u_\lambda$, so $\psi(a) \in \mathcal{L} \subset K_B$ by 3.3.

We proved $\psi(K_A) \subseteq K_B$. The converse inclusion follows in the same way so we have $\psi(K_A) = K_B$. Now K_A, K_B are dense in A,B respectively and $\psi | M(A)$ is an isometry so we have $\psi(A) = B$.

CHAPTER 8. IDEALS AND ORDER IN $\Gamma(K)$

In the first part of this chapter we shall discuss the κ-closed two-sided ideals of $\Gamma(K)$. The second part comprises some results on the partial order induced in $\Gamma(K)$ be the convex cone $\Gamma(K)^+$. As before, A denotes an arbitrary C*-algebra and K its Pedersen ideal.

8.1. <u>Proposition</u>. There is a one to one correspondence between the family of κ-closed two-sided ideals of $\Gamma(K)$ and the norm closed two-sided ideals of A given by $J \to J \cap A$. The norm closure of $J \cap K$ is $J \cap A$; the κ-closure of $J \cap K$ is J.

<u>Proof</u>. Clearly $J \cap A$ is a norm closed two-sided ideal of A. Now let I be a norm closed two-sided ideal and let J be the κ-closure of I. Since multiplication is separately continuous in the κ-topology and K is κ-dense in $\Gamma(K)$, it follows that J is a two-sided ideal of $\Gamma(K)$. Let $x \in J \cap A$ and $\{e_\lambda\}$ be an approximate identity for A contained in K. Then clearly $xe_\lambda \in I$ which means $x \in I$, so $J \cap A = I$ and our first assertion holds.

We now wish to prove the last two assertions. Obviously, the norm (κ-) closure of $J \cap K$ is included in $J \cap A$ (J). Let $\{e_\lambda\}$ be an approximate identity of A contained in K. If $x \in J$, then $x = \kappa\text{-lim}_\lambda xe_\lambda$; if $x \in J \cap A$, then x is the norm limit of $\{xe_\lambda\}$. In both cases, $\{xe_\lambda\} \subseteq J \cap K$.

8.2. <u>Proposition</u>. A κ-closed two-sided ideal of $\Gamma(K)$ is positively generated.

<u>Proof</u>. Let J be a κ-closed two-sided ideal of $\Gamma(K)$. Then $J \cap A$ is self-adjoint, so its κ-closure J is self-adjoint too. Define

$$f^+(t) = \max(t,0) \text{ and } f^-(t) = \max(-t,0).$$

For each $x \in J$ with $x^* = x$, the elements $x^+ = f^+(x)$, $x^- = f^-(x)$ are in $J \cap \Gamma(K)^+$ by 5.17, 5.19 and 5.23. Clearly, $x = x^+ - x^-$, thus J is the linear span of J^+.

8.3. <u>Proposition</u>. Let J_1, J_2 be κ-closed two-sided ideals of $\Gamma(K_A)$, $J = J_1 \cap J_2$, $I = J \cap A$. If the canonical map $\phi:\Gamma(K_A) \to \Gamma(K_{A/I})$ is onto, then $J_1 + J_2$ is κ-closed.

<u>Proof</u>. Let $\{x_\alpha^1\}$, $\{x_\alpha^2\}$ be nets in J_1, J_2 respectively, such that $\{x_\alpha^1 + x_\alpha^2\}$ κ-converges to some $y \in \Gamma(K_A)$. Then for each $a \in K$,

$$ya = \lim_\alpha (x_\alpha^1 a + x_\alpha^2 a),$$

so $ya \in I_1 + I_2$ by [13, 1.8.4, p. 18], where $I_k = J_k \cap A$. Thus

$$\phi(ya) \in \phi(I_1) + \phi(I_2).$$

Since $\phi(I_1) \cap \phi(I_2) = \{0\}$ and $K_{A/I} = \phi(K_A)$, we can define $S_1, S_2 : K_{A/I} \to A/I$ by

$$(*) \quad \phi(ya) = S_1(\phi(a)) + S_2(\phi(a)),\ S_k(\phi(a)) \in \phi(I_k).$$

Similarly, we can define $T_1, T_2 : K_{A/I} \to A/I$ by

$$({}^*_*) \quad \phi(ay) = T_1(\phi(a)) + T_2(\phi(a)),\ T_k(\phi(a)) \in \phi(I_k).$$

It is obvious that these maps are well defined and satisfy

$$\phi(b)\, S_k(\phi(a)) = T_k(\phi(b))\, \phi(a),$$

for each $a,b \in K_A$. By 3.5,

$$(S_k, T_k) \in \Gamma(K_{A/I}).$$

By virtue of our hypothesis, there are $x_1, x_2 \in \Gamma(K_A)$ such that $\phi(x_k) = (S_k, T_k)$. Now, for each $a \in K$,

$$\phi(x_k a) \in \phi(I_k),$$

which means

$$x_k a \in I_k + I = I_k.$$

Thus $x_k \in J_k$ and

$$\phi(y) = \phi(x_1 + x_2)$$

by (*) and $(\overset{*}{*})$. Clearly,

$$z = y - (x_1 + x_2) \in J,$$

so y

$$y = x_1 + (x_2 + z)$$

is in $J_1 + J_2$.

The following example shows $J_1 + J_2$ fails to be κ-closed unless some restrictions are imposed.

8.4. <u>Example</u>. Let ω be the first infinite ordinal and Ω the first uncountable ordinal. Consider the segments of ordinals $[0,\Omega]$, $[0,\omega]$ with the order topology,

$$X' = [0,\Omega] \times [0,\omega]$$

with the product topology and

$$X = X' \setminus \{(\Omega,\omega)\}.$$

Let

$$F_1 = \{(\alpha,\omega) = 0 \leq \alpha < \Omega\}$$

$$F_2 = \{(\Omega,\beta) = 0 \leq \beta < \omega\}$$

and

$$J_k = \{f \in C(X) : f_{|F_k} = 0\}.$$

Then J_1, J_2 are κ-closed ideals of $C(X)$, but $J_1 + J_2$ is not κ-closed. Indeed, it is easy to see that for each $f \in J_2$ there exists an ordinal α, $0 \leq \alpha < \Omega$, such that $f(\alpha',\omega) = 0$ for each $\alpha' \geq \alpha$. Hence the same condition is satisfied by each $f \in J_1 + J_2$. Thus $1_X \notin J_1 + J_2$. Now, for every $\alpha < \Omega$ and $\beta < \omega$ define

$$\chi_{\alpha,\beta} \in C(X)$$

by

$$\chi_{\alpha,\beta}(\alpha',\beta') = \begin{cases} 0, & \alpha < \alpha', \beta < \beta' \\ 1, & \text{otherwise}. \end{cases}$$

Obviously, $\{\chi_{(\alpha,\beta)}\}$ converges to 1_X uniformly on each compact subset of X. Let

$$\chi^1_{\alpha,\beta}(\alpha',\beta') = \begin{cases} 0, & \beta < \beta' \\ 1, & \text{otherwise} \end{cases}$$

and $\chi^2_{\alpha,\beta} = \chi_{\alpha,\beta} - \chi^1_{\alpha,\beta}$. Then $\chi^k_{\alpha,\beta} \in J_k$, so $\chi_{\alpha,\beta} \in J_1 + J_2$. Thus 1_X belongs to the κ-closure of $J_1 + J_2$.

8.5. Proposition. $\Gamma(K)^+$ is κ-closed.

Proof. Let $\{x_\alpha\}$ be a net in $\Gamma(K)^+$ which κ-converges to some $x \in \Gamma(K)$. Clearly $x^* = x$, so $x = x^+ - x^-$ with $x^+, x^- \in \Gamma(K)^+$, $x^+ \cdot x^- = 0$. For each $\pi \in \hat{A}$, the net $\{\pi(x_\alpha)\}$ strongly converges to $\pi(x)$, since

$$\pi(K)[H_\pi] = H_\pi,$$

therefore, $\pi(x) \geq 0$. From

$$\pi(x^+)\pi(x^-) = 0,$$

$$\pi(x) = \pi(x^+) - \pi(x^-),$$

and

$$\pi(x^+),\ \pi(x^-) \geq 0$$

we infer $\pi(x^-) = 0$. Thus, $\pi(x^-) = 0$ for every $\pi \in A$, hence $x^- = 0$ and $x \in \Gamma(K)^+$.

We shall prove now a Riesz decomposition property for $\Gamma(K)$ by using the ideas of [29] where this property was established for A.

8.6. Lemma. Let $\{x_n\}_{n=1}^\infty$, $\{y_n\}_{n=1}^\infty$ be two sequences in $\Gamma(K)$ which κ-converge to x,y respectively. Then

$$xy = \kappa\text{-}\lim_{n\to\infty} x_n y_n.$$

Proof. Let $a \in K^+$ and $\epsilon > 0$. By virtue of 3.4 and the uniform boundedness theorem there is a number $M > 0$ such that

$$\|x_n za\| \leq M\|za\|$$

for every $z \in \Gamma(K)$ and $n = 1,2,\ldots$. Let N be a natural number such that for $n > N$ we have

$$\|x_n ya - xya\| < \frac{1}{2}\epsilon,$$

$$\|y_n a - ya\| < \epsilon(2M)^{-1}.$$

Then

$$\|x_n y_n a - xya\| \leq \|x_n(y_n a - ya)\| + \|(x_n - x)ya\| < \epsilon$$

if $n > N$. One proves in a similar way that

$$\lim_{n\to\infty} \|axy - ax_n y_n\| = 0.$$

8.7. Proposition. Let

$$\{x_i\}_{i=1}^n \subseteq \Gamma(K)^+,$$

$y \in \Gamma(K)$. If

$$0 \leq y \leq \Sigma_{i=1}^n x_i,$$

then there exist $y_i \in \Gamma(K)$, $1 \leq i \leq n$ such that

$$y = \Sigma_{i=1}^n y_i^* y_i$$

and

$$y_i y_i^* \leq x_i, \quad 1 \leq i \leq n.$$

Proof. Set $x = \Sigma_{i=1}^n x_i$ and $u_p = (p^{-1} + x)^{-1}x$ for each natural number p. We shall show that

$$y^{1/2} = \kappa\text{-}\lim_{p\to\infty} y^{1/2}u_p.$$

Let $\pi \in \hat{A}$. Then we have

$$\|\pi(y^{1/2} - y^{1/2}u_p)\|^2 = \|(\pi(1) - \pi(u_p))\pi(y)(\pi(1) - \pi(u_p))\| \leq$$

$$\leq \|(\pi(1) - \pi(u_p))\pi(x)(\pi(1) - \pi(u_p))\| \leq$$

$$\leq \sup_{t\geq 0} [t^{1/2} - (p^{-1} + t)^{-1}t^{3/2}]^2 = (4p)^{-1}.$$

Hence, for each $a \in K$, we have

$$\|(y^{1/2} - y^{1/2}u_p)a\| =$$

$$= \sup \{\|\pi((y^{1/2} - y^{1/2}u_p))\pi(a)\| : \pi \in \hat{A}\} \leq 2^{-1}p^{-1/2}\|a\|,$$

and this proves the claim. One shows in the same way that

$$y^{1/2} = \kappa\text{-}\lim_{p\to\infty} u_p y^{1/2}.$$

By 8.6,

$$y = \kappa\text{-}\lim_{p\to\infty} y^{1/2}u_p^2 y^{1/2} =$$

$$= \kappa\text{-}\lim_{p\to\infty} \Sigma_{i=1}^n (y^{1/2}(p^{-1}+x)^{-1}x^{1/2}x_i^{1/2})(x_i^{1/2}x^{1/2}(p^{-1}+x)^{-1}y^{1/2}).$$

We claim now that for each i, $1 \leq i \leq n$, the sequence

$$\{x_i^{1/2}\ x^{1/2}\ (p^{-1} + x)^{-1}\ y^{1/2}\}_{p=1}^{\infty}$$

is κ-Cauchy. Denote

$$y_{ip} = x_i^{1/2}\ x^{1/2}\ (p^{-1} + x)^{-1}\ y^{1/2}.$$

For each $\pi \in \hat{A}$, the following holds:

$$\|\pi(y_{ip} - y_{iq})\|^2 = \|\pi[x_i^{1/2} x^{1/2}((p^{-1} + x)^{-1} - (q^{-1} + x)^{-1}y^{1/2}]\|^2 \leq$$

$$\leq \|\pi[y^{1/2}((p^{-1} + x)^{-1} - (q^{-1} + x)^{-1})^2 x^2 y^{1/2}]\| =$$

$$= \|\pi((u_p - u_q)y^{1/2})\|^2.$$

As above, it follows that for each $a \in K$,

$$\lim_{\substack{p\to\infty \\ q\to\infty}} (y_{ip} - y_{iq})a = \lim_{\substack{p\to\infty \\ q\to\infty}} a(y_{ip} - y_{iq}) = 0.$$

Set

$$y_i = \kappa\text{-}\lim_{p\to\infty} y_{ip}, \quad 1 \leq i \leq n.$$

Again, by 8.6, we get

$$y = \Sigma_{i=1}^n y_i^* y_i.$$

Since

$$y_{ip}\, y_{ip}^* = x_i^{1/2}\, x^{1/2}(p^{-1} + x)^{-1} y(p^{-1} + x)^{-1}\, x^{1/2}\, x_i^{1/2} \leq$$

$$\leq x_i^{1/2}\, u_p^2\, x_i^{1/2} \leq x_i,$$

8.5 and 8.6 imply

$$y_i\, y_i^* \leq x_i, \quad 1 \leq i \leq n.$$

8.8. <u>Proposition</u>. Let J be a κ-closed two-sided ideal of $\Gamma(K)$. If $0 \leq x \leq y$, $y \in J$, then $x \in J$.

Proof. Let $\{e_\lambda\}$ be a bounded approximate identity for A contained in K. For each λ we have

$$0 \leq e_\lambda^* x e_\lambda \leq e_\lambda^* y e_\lambda, \quad e_\lambda^* y e_\lambda \in J \cap A.$$

Since $J \cap A$ is a uniformly closed two-sided ideal of A,

$$e_\lambda^* x e_\lambda \in J \cap A.$$

Thus,

$$x = \kappa\text{-}\lim_\lambda e_\lambda^* x e_\lambda$$

is in J.

As has been done in [29] we shall use the decomposition property in order to derive the following:

8.9. Proposition. If J_1, J_2 are κ-closed two-sided ideals of A, then

$$(J_1 + J_2)^+ = J_1^+ + J_2^+.$$

Proof. Clearly,

$$J_1^+ + J_2^+ \subseteq (J_1 + J_2)^+.$$

Let

$$x \in (J_1 + J_2)^+.$$

Then $x = y_1 + y_2$ with $y_k = y_k^*$, $y_k \in J_k$, $k = 1,2$. Thus

$$x \leq |y_1| + |y_2|.$$

By 8.7, there exist $u_1, u_2 \in \Gamma(K)$ such that

$$x = u_1^* u_1 + u_2^* u_2,$$

$$u_k u_k^* \leq |y_k|.$$

But $|y_k| \in J_k$, so by virtue of 8.8,

$$u_k u_k^* \in J_k.$$

Thus,

$$u_k^*(u_k u_k^*)u_k = (u_k^* u_k)^2 \in J_k,$$

hence,

$$u_k^* u_k \in J_k$$

by virtue of 5.17.

CHAPTER 9. DERIVATIONS ON SUBALGEBRAS of $\Gamma(K)$

Throughout this section A will denote a C*-algebra and K will denote its Pedersen ideal. In this chapter we shall study derivations on subalgebras of $\Gamma(K)$.

9.1. Definitions. Let B be an algebra and let δ be a linear mapping of B into B. If $\delta(xy) = \delta(x)y + x\delta(y)$ for each c,y in B, then δ is called a derivation on B. If there exists an element a in B such that

$$\delta(x) = [a,x] \equiv ax - xa$$

for each $x \in B$, then δ is said to be inner. Otherwise δ is called an outer derivation.

9.2. Lemma. Let B be a self-adjoint subalgebra of $\Gamma(K)$, ϕ a representation of A, $\overline{\Phi}$ the induced representation of $\Gamma(K_A)$, and δ a derivation on B. If B is either κ-closed or equal to K and $\overline{\phi}(x) = \overline{\phi}(y)$ for some x,y in B, then

$$\overline{\phi}(\delta(x)) = \overline{\phi}(\delta(y)).$$

Consequently, the map

$$\tilde{\delta}:\overline{\phi}(B) \to \overline{\phi}(B)$$

defined by $\delta(\overline{\phi}(x)) = \overline{\phi}(\delta(x))$ is a derivation on $\overline{\phi}(B)$.

Proof. Suppose $\overline{\phi}(x) = \overline{\phi}(y)$ for x,y in B. Clearly, we may assume x and y are hermitian. Let $z = x - y$. First, let us assume that B is κ-closed. Since $\overline{\phi}$ is continuous under the κ_A and $\kappa_{\phi(A)}$ topologies, respectively, $(\ker \overline{\phi}) \cap B$ is a κ_A-closed, self-adjoint subalgebra of $\Gamma(K)$ that contains z. It follows from 5.21 that there exist elements z_1, z_2 in $(\ker \overline{\phi}) \cap B$ such that $z = z_1 z_2$. Therefore,

$$\overline{\phi}(\delta(x - y)) = \overline{\phi}(\delta(z_1 z_2)) = \overline{\phi}(\delta(z_1)z_2 + z_1\delta(z_2)) = 0.$$

Now assume $B = K$. Let D be the C*-algebra of A generated by z. By 2.6, $D \subset K$, and since $z \in \ker \phi$,

$$D \subset K \cap (\ker \phi).$$

The rest of the proof proceeds as before.

9.3. <u>Proposition</u>. Let B be a κ_A-closed, self-adjoint commutative subalgebra of $\Gamma(K)$. If δ is a derivation on B, then $\delta = 0$.

<u>Proof</u>. Let π be an irreducible representation of A and let δ be the derivation on $\overline{\pi}(B)$ given in 9.2. By virtue of 5.21, 4.4, and 5.18, it is easy to see that

$$\overline{\pi}(B) = \overline{\pi}(B \cap M(A)).$$

Hence, $\overline{\pi}(B)$ is a C*-algebra, since $B \cap M(A)$ is a C*-algebra. It is well known that the only derivation on a commutative C*-algebra is the zero derivation. Therefore,

$$\delta(\overline{\pi}(x)) = \overline{\pi}(\delta(x)) = 0$$

for each $x \in B$. Since π was arbitrarily chosen, $\delta = 0$.

Now let δ be a derivation on K, π an irreducible representation of A, and $a \in K$. For each $h \in H_\pi$, $h \neq 0$, let the map

$$\hat{\delta}_h : \mathcal{R}_{a*} \to H_\pi$$

be defined by

$$\hat{\delta}_h(x) = \pi(\delta(x))(h)$$

(see 3.3).

The proof of the next lemma is a variant of an argument given in [23]. However, because of our different setting, we shall include the proof for the convenience of the reader.

9.4. Lemma. For each $h \in H_\pi$ the linear map $\hat{\delta}_h$ is bounded.

Proof. First, we will show that if $\hat{\delta}_h$ is bounded for some $h \in H_\pi$, then $\hat{\delta}_h$ is bounded for all $h \in H_\pi$. Let $h_0 \in H_\pi$ and suppose $\hat{\delta}_{h_0}$ is bounded. Now let $h \in H_\pi$. Since π is an irreducible representation, h_0 is a strictly cyclic vector of H_π. Consequently, there is a $b \in K$ such that

$$h = \pi(b)(h_0).$$

Since

$$\hat{\delta}_h(x) = \pi(\delta(x))(h) = \pi(\delta(x))(\pi(b)(h_0)) =$$

$$\pi(\delta(x)b)(h_0) = \pi(\delta(xb))(h_0) - \pi(x\delta(b))(h_0)$$

$$= \hat{\delta}_{h_0}(xb) - \pi(x)\pi(\delta(b))(h_0),$$

it is clear that $\hat{\delta}_h$ is bounded.

So assume that our assertion is false. This means that $\hat{\delta}_h$ is discontinuous for each $h \in H_\pi$. Since any derivation on a C*-algebra is bounded and

$$\pi(K) = K_{\pi(A)} = \pi(A)$$

when H_π is of finite dimension, it follows from 9.2 that H_π must be of infinite dimension. Choose an $h \in H_\pi$, $h \neq 0$. Since π is irreducible and H_π is infinite dimensional, we can find a sequence of elements $\{b_n\}_{n=0}^{\infty}$ in K so that

$$\{\pi(b_n)(h)\}_{n=0}^{\infty}$$

is a sequence of linearly independent vectors in H_π with

$$\|\pi(b_n)(h)\| = 1.$$

By virtue of [13, 2.8.2, p. 43], we can choose $w_1, w_2, \ldots$ in A by induction such that the following hold:

$$\pi(w_i b_0)(h) = \pi(w_i b_1)(h) = \ldots = \pi(w_i b_{i-1})(h) = 0;$$

$$\pi(w_i b_i)(h)$$

is not a linear combination of

$$\pi(v_i b_1)(h), \pi(v_i b_2)(h), \ldots, \pi(v_i b_i)(h),$$

where

$$v_i = w_1 + \ldots + w_{i-1}; \|w_i\| < 2^{-i}.$$

Now set

$$v = \Sigma_{i=1}^{\infty} w_i.$$

Note

$$\pi(vb_0)(h) = 0$$

and

$$\pi(vb_i)(h) = \pi(v_i b_i)(h) + \pi(w_i b_i)(h)$$

which is not a linear combination of

$$\pi(vb_1)(h),\ \pi(vb_2)(h),\ldots,\pi(vb_{i-1})(h),$$

since

$$\pi(vb_j)(h) = \pi(v_i b_j)(h)$$

for $1 \leq j < i$. Hence

$$\pi(vb_1)(h),\ \pi(vb_2)(h),\ \ldots$$

are linearly independent.

By virtue of what we have just shown we can choose by induction a sequence $y_1, y_2, \ldots$ in A such that for each integer $n > 0$ the following hold:

(i) $\pi(y_n y_{n-1} \cdots y_1 b_{n-1})(h) = 0$;

(ii) $\{\pi(y_n y_{n-1} \cdots y_1 b_j)(h)\}_{j=n}^{\infty}$ is linearly independent;

(iii) $\|y_n\| \leq 1/2^n$.

Now we can choose by induction using the discontinuity of all $\hat{\delta}_n$, a sequence $x_1, x_2, \ldots$ in $\mathfrak{A}_{a*}$ such that

$$\text{(a)}\ \|x_n\| \leq \min\left\{\frac{1}{1+\|\pi(\delta(y_j\cdots y_1 b_i))\|\,\|h\|},\ \frac{1}{1+\|\pi(\delta(b_i))\|\,\|h\|} :\right.$$

$$\left. 1 \leq i \leq n,\ 1 \leq j \leq n\right\};$$

$$\text{(b)}\ \|\pi(\delta(x_n)y_n\cdots y_1 b_n)(h)\| \geq n. +$$

$$+ \Sigma_{j=1}^{n-1} \|\pi(\delta(x_j y_j \cdots\cdots\cdots y_1)b_n)(h)\|.$$

Finally, we put

$$z = \Sigma_{j=1}^{\infty} x_j y_j \cdots y_1$$

and

$$z_n = x_{n+1} + \Sigma_{j=n+2}^{\infty} x_j y_j \cdots y_{n+2},$$

the series converging absolutely by (iii) and (a). Note that z and z_n are in $\mathcal{R}_{a*}$, since $\mathcal{R}_{a*}$ is a right ideal. Also, note that

$$z = \Sigma_{j=1}^{n-1} x_j y_j \cdots y_1 + x_n y_n \cdots y_1 + z_n y_{n+1} \cdots y_1.$$

So

$$\|\delta(z)\| \geq \|\pi(\delta(z)b_n)(h)\|$$

$$\geq \|\pi(\delta(\Sigma_{j=1}^{n-1} x_j y_j \cdots y_1)b_n + \delta(x_n y_n \cdots y_1)b_n$$

$$+ \delta(z_n y_{n+1} \cdots y_1)b_n)(h)\|$$

$$\geq \|\pi(\delta(x_n)y_n \cdots y_1 b_n + x_n \delta(y_n \cdots y_1 b_n) - x_n y_n \cdots y_1 \delta(b_n)$$

$$+ \delta(z_n)y_{n+1} \cdots y_1 b_n + z_n \delta(y_{n+1} \cdots y_1 b_n)$$

$$- z_n y_{n+1} \cdots y_1 \delta(b_n) + \Sigma_{j=1}^{n-1} \delta(x_j y_j \cdots y_1)b_n)(h)\|$$

$$\geq \|\pi(\delta(x_n)y_n\cdots y_1b_n)(h)\| - \|x_n\|\|\pi(\delta(y_n\cdots y_1b_n))(h)\|$$

$$- \|x_n\|\|\pi(\delta(b_n))(h)\| - \|\pi(\delta(z_n)y_{n+1}\cdots y_1b_n)(h)\|$$

$$- \|z_n\|\|\pi(\delta(y_{n+1}\cdots y_1b_n))(h)\| - \|z_n\|\|\pi(\delta(b_n))(h)\|$$

$$- \Sigma_{j=1}^{n-1} \|\pi(\delta(x_jy_j\cdots y_1)b_n)(h)\|.$$

By (b),

$$\|\pi(\delta(x_n)y_n\cdots y_1b_n)(h)\| - \Sigma_{j=1}^{n-1} \|\pi(\delta(x_jy_j\cdots y_1)b_n)(h)\| \geq n,$$

by (a),

$$\|x_n\|\|\pi(\delta(y_n\cdots y_1b_n))(h)\| \leq 1$$

and

$$\|x_n\|\|\pi(\delta(b_n))(h)\| \leq 1,$$

by (i),

$$\|\pi(\delta(z_n)y_{n+1}\cdots y_1b_n)(h)\| = 0,$$

and by (a) and (iii),

$$\|z_n\|\|\pi(\delta(y_{n+1}\cdots y_1b_n))(h)\| \leq (\|x_{n+1}\| +$$

$$+ \Sigma_{j=n+2}^{\infty}(\|x_j\|/2^j)\|\pi(\delta(y_{n+1}\cdots y_1b_n))(h)\| \leq 2$$

and similarly,

$$\|z_n\|\|\pi(\delta(b_n))(h)\| \leq 2.$$

Hence

$$\|\delta(z)\| \geq n-6$$

which is a contradiction so $\hat{\delta}_h$ must be continuous.

9.5. Theorem. For each derivation δ on K there is a unique derivation δ on $\Gamma(K)$ that extends δ; moreover, δ is κ-continuous.

Proof. Let $a \in K$. Define the linear map $\rho_a : \mathcal{R}_{a*} \to A$ by the formula

$$\rho_a(x) = \delta(x),$$

$x \in \mathcal{R}_{a*}$. We will now show that the linear map ρ_a is bounded by using the closed graph theorem. Let $\{x_n\}$ be a sequence in $\mathcal{R}_{a*}$ that converges to x and suppose $\rho_a(x_n)$ converges to y. To show $\rho_a(x) = y$ it will suffice to show

$$\pi(\rho_a(x))(h) = \pi(y)(h)$$

for each irreducible representation π of A and $h \in H_\pi$. But by virtue of 9.4,

$$\pi(\rho_a(x))(h) = \delta_h(x) = \lim \delta_h(x_n) = \lim \pi(\rho_a(x_n))(h) = \pi(y)(h).$$

Hence, $\rho_a(x) = y$ and ρ_a is bounded. Since the map $x \to \delta^*(x)$ must also be bounded on $\mathcal{R}_{a*}$, where $\delta^*(x) = \delta(x^*)^*$, we see that the map

$$\lambda_a : \mathcal{L}_a \to A$$

defined by

$$\lambda_a(x) = \delta(x)$$

is also bounded.

Now let $x \in \Gamma(K)$. Since K is κ-dense in $\Gamma(K)$, there exists a net $\{x_\alpha\}$ in K that converges to x. Since

$$\delta(x_\alpha)a = \delta(x_\alpha a) - x_\alpha\delta(a) = \lambda_a(x_\alpha a) - x_\alpha\delta(a)$$

and

$$a\delta(x_\alpha) = \rho_a(ax_\alpha) - \delta(a)x_\alpha$$

for each $a \in K$, $\{\delta(x_\alpha)\}$ is a κ-Cauchy net in K and therefore has a κ-limit $\tilde{\delta}(x)$ in $\Gamma(K)$. Clearly, $\delta(x)$ is independent of the net chosen converging to x. It is also clear from the above that the map $\tilde{\delta}:\Gamma(K) \to \Gamma(K)$ is κ-continuous. Since multiplication in $\Gamma(K)$ is separately continuous, $\tilde{\delta}$ is a derivation on $\Gamma(K)$.

9.6. Corollary. Each derivation δ on $\Gamma(K)$ is κ-continuous.

Proof. Let $a \in K$ and B_a the C*-algebra generated by a. Since $B_a \subset K$, we have by the Cohen-Hewitt factorization theorem elements b,c in K such that $a = bc$. Hence

$$\delta(a) = b\delta(c) + \delta(b)c \in K$$

and therefore K is invariant under δ. The conclusion now follows from 9.5.

We will now view A as a sub C*-algebra of B(H) for some fixed Hilbert space H. Moreover, we will assume $A[H] = H$. Recall that $\Gamma(K)$ can be viewed as the *-algebra of all operators T acting in the dense subspace K[H] which have the property that $xT + Ty$ is bounded on K[H] for each x,y in K and its unique extension belongs to K. Now let D(K) denote all of the linear maps $T:K[H] \to K[H]$ such that Tx, xT are bounded and the unique extension of $Tx - xT$ belongs to K for all $x \in K$. Clearly $D(K) \subseteq \Gamma(K)$ and the map $x \to [T,x]$ is a derivation on K. Moreover, this characterizes the derivations on K as we shall see in the following theorem.

9.7. Theorem. Let δ be a derivation on K. Then there exists a $T \in D(K)$ such that $\delta(x) = [T,x]$ for each $x \in K$. Moreover, if A is an ideal of a W*-algebra, then $D(K) = \Gamma(K)$.

Before we can proceed with the proof we first need some lemmas. Let

$$D = \mathcal{L}_a \cap \mathcal{R}_a$$

for some fixed a in K^+, and $\delta_a = \delta | D$. Note that from the proof of 9.5, we see that δ is bounded. Let $\overline{D}$ and $\overline{A}$ denote the closure of D and A in the weak operator topology.

9.8. Lemma. There exists a linear map $\tilde{\delta}_a : \overline{D} \to \overline{A}$ that extends δ_a,

$$\|\tilde{\delta}_a\| = \|\delta_a\| ,$$

and is continuous in the ultraweak topology; moreover,

$$\tilde{\delta}_a(xy) = x\tilde{\delta}_a(y) + \tilde{\delta}_a(x)y$$

for each x,y in $\overline{D}$.

Proof. The proof is identical to [12, Lemma 4, p. 309].

Let [a] be the support projection of a. Of course $[a] \in \overline{D}$ and it is easy to verify $[a]\overline{A}[a] = \overline{D}$. Note [a] is the identity for $\overline{D}$. So the map

$$\delta_{[a]} : \overline{D} \to \overline{D}$$

defined by

$$\delta_{[a]}(x) = [a]\tilde{\delta}_a(x)[a]$$

is a derivation on $\overline{D}$. It follows that for each projection e in $\overline{D}$ the map

$$\delta_e : e\bar{D}e \to e\bar{D}e$$

defined by

$$\delta_e(x) = e\delta_{[a]}(x)e$$

is a derivation on $e\bar{D}e$. Recall that in 9.5 we showed that the maps

$$\lambda_b : \mathcal{L}_b \to A$$

and

$$\rho_b : \mathcal{R}_b \to A$$

were bounded.

The proof of the next lemma is a variant of the argument given for [35, 4.1.6, p. 156].

9.9. Lemma. Let e be any projection in $\bar{D}$. Then there exists an element x_e in $e\bar{D}e$ that satisfies the following:

(i) $\delta_e(y) = [x_e, y]$ for each $y \in e\bar{D}e$;

(ii) for each $b \in K$,

$$\|x_e b\| \leq 4\|\lambda_b\|\,\|b\| \text{ and } \|bx_e\| \leq 4\|\rho_b\|\,\|b\|;$$

(iii) $\|x_e\| \leq \|\delta_e\|$.

Proof. Set $B = e\bar{D}e$ and let B^u be the group of all unitary elements in B. For $u \in B^u$, put

$$T_u(x) = (ux + \delta_e(u))u^{-1},$$

$x \in B$. Note that if $u, v \in B$, then

$$T_u T_v(x) = T_{uv}(x)$$

(see [35, 4.1.6, p. 156]). Let Δ be the collection of all convex subsets S of B that are closed in the ultraweak topology and satisfy the following conditions:

(a) $T_u(S) \subseteq S$, $u \in B^u$;

(b) $\|xb\| \leq 2\|\lambda_b\|\|b\|$ and $\|bx\| \leq 2\|\rho_b\|\|b\|$ for each $x \in S$ and $b \in K$;

(c) $\|x\| \leq \|\delta_e\|$, $x \in S$. Now we will show that Δ is non-empty.

Let S be the closure of the convex hull of $\{T_u(0) | u \in B^u\}$ in the ultraweak topology. Since

$$T_u T_v(0) = T_{uv}(0) \text{ and } \|T_u(0)\| = \|\delta_e(u)u^{-1}\|,$$

we see that (a) and (c) hold for the convex hull of

$$\{T_u(0) | u \in B^u\}$$

and consequently for S. Due to the fact that S is also the closure of the convex hull of

$$\{T_u(0) | u \in B^u\}$$

in the strong operator topology [12, Théorèm 1, p. 38], we need only to show that (b) holds for

$$\{T_u(0) | u \in B^u\}.$$

First, note

$$\|T_u(0)b\| = \|\delta_e(u)u^{-1}b\| = \|u\delta_e(u^{-1})b\| \leq \|\tilde{\delta}_a(u^{-1})eb\|.$$

Now since the unit ball of D is ultraweakly dense in the unit ball of $\overline{D}$, there exist nets $\{x_\alpha\}$ and $\{y_\beta\}$ in the unit ball of D that converge ultraweakly to u^{-1} and e respectively. So for h_1, h_2 in the unit ball of H

$$\begin{aligned}
\|\langle\tilde{\delta}_a(u^{-1})eb(h_1),h_2\rangle| &= \lim_\alpha \lim_\beta |\langle\delta(x_\alpha)y_\beta b(h_1),h_2\rangle| \\
&\leq \overline{\lim_{\alpha,\beta}} |\langle\delta(x_\alpha y_\beta b)(h_1),h_2\rangle| \\
&+ \overline{\lim_{\alpha,\beta}} |\langle x_\alpha\delta(y_\beta b)(h_1),h_2\rangle| \\
&\leq \overline{\lim_{\alpha,\beta}} \|\delta(x_\alpha y_\beta b)\| + \overline{\lim_{\alpha,\beta}} \|\delta(y_\beta b)\| \\
&= \overline{\lim_{\alpha,\beta}} \|\lambda_b(x_\alpha y_\beta b)| + \overline{\lim_{\alpha,\beta}} \|\lambda_b(y_\beta b)\| \\
&\leq 2\|\lambda_b\|\|b\|.
\end{aligned}$$

Hence,

$$\|T_u(0)b\| \leq 2\|\lambda_b\|\|b\|.$$

Similarly,

$$\|bT_u(0)\| \leq 2\|\rho_b\|\|b\|$$

and therefore Δ is non-empty.

Just as in [35, 4.1.6, p. 156], there exists a minimal element S_0 in Δ by Zorn's lemma. By an argument identical to the proof of [35, 4.1.6, p. 156], we can show that S_0 consists of one element x_e whenever B is either semi-finite or purely infinite. But for this case

$$T_u(x_e) = ux_e u^{-1} + \delta_e(u)u^{-1} = x_e,$$

or equivalently,

$$\delta_e(u) = x_e u - u x_e .$$

Since any element of B can be written as a linear combination of unitary elements,

$$\delta_e(y) = [x_e, y], \; y \in B.$$

Observe that x_e satisfies (i) and (iii) and that

$$\|x_e b\| \leq 2\|\lambda_b\| \|b\| \text{ and } \|b x_e\| \leq 2\|\rho_b\| \|b\| .$$

Since $B = B_1 \oplus B_2$ with B_1 semi-finite and B_2 purely infinite, we can reduce the proof to B_1 and B_2. So (ii) holds and our proof is complete.

9.10. Proof of theorem. For each $a \in K^+$ let $[a]$ denote the support projection of a and $x_{[a]}$ the element given by 9.9 satisfying the conditions listed. Note that $x_{[a]}$ is a κ-bounded net where the order of the directed set is the usual one for projections. Consequently, the set

$$Q_{(h_1;h_2)} = \{\lambda : |\lambda| \leq \sup_{a \in K} \{|\langle h_1, x_{[a]}(h_2)\rangle|\},$$

$h_1 h_2 \in K[H]$ is a compact subset of the plane. Let $x^0_{[a]}$ be the element of the product space $\pi Q_{(h_1;h_2)}$ given by

$$x^0_{[a]}(h_1, h_2) = \langle h_1, x_{[a]}(h_2)\rangle.$$

Since by Tychonoff's theorem

$$\pi Q_{(h_1;h_2)}$$

is compact, we may assume by dropping to a subset if necessary that $x^0_{[a]}$ converges in the product topology to some

$$x^0 \in \pi Q_{(h_1;h_2)} .$$

For

$$h_0 \in K[H]$$

define the linear functional f on K[H] by

$$f(h) = \lim \langle h, x_{[a]}(h_0) \rangle .$$

Since $\{x_{[a]}\}$ is κ-bounded, there exists an $T(h_0) \in H$ such that

$$\lim \langle h, x_{[a]}(h_0) \rangle = \langle h, T(h_0) \rangle$$

for all $h \in H$. Clearly, the map $h \to T(h)$ is a linear map from K[H] into H. But for $b \in K$ and $h_1, h_2 \in K[H]$,

$$\langle Tb(h_1), h_2 \rangle = \lim \langle x_{[a]} b(h_1), h_2 \rangle$$

$$= \lim \langle x_{[a]} b - b x_{[a]}(h_1), h_2 \rangle + \lim \langle b x_{[a]}(h_1), h_2 \rangle$$

$$= \lim \ \langle \delta_{[a]}(b)(h_1), h_2 \rangle + \langle bT(h_1), h_2 \rangle$$

$$= \lim \langle [a] \delta(b) [a](h_1), h_2 \rangle + \langle bT(h_1), h_2 \rangle$$

$$= \langle (\delta(b) + bT)(h_1), h_2 \rangle .$$

This means

$$Tb(h) = \delta(b)(h) + bT(h)$$

for each $h \in K[H]$, so $T: K[H] \to K[H]$ and $Tb - bT = \delta(b) \in K$ for each $b \in K$. From the above it is also clear that Tb and bT are bounded. Hence $T \in D(K)$ and $\delta(b) = [T, b]$.

Now suppose A is an ideal in a W*-algebra. Then by [37, Corollary 2.2, p. 478], M(A) is a W*-algebra. It follows by the way each $x_{[a]}$ is defined that $x_{[a]} \in M(A)$. To show that $T \in \Gamma(K)$, we need only show that $Tb \in K$ or, since b can be factored, $Tb \in M(A)$ for each $b \in K^+$. But

$$\langle Tb(h),h\rangle = \lim \langle x_{[a]}b(h),h\rangle \ .$$

So $\{x_{[a]}b\}$ is a bounded net that converges to Tb in the weak operator topology. Thus $Tb \in M(A)$ and our proof is complete.

This next result was suggested by [1, Theorem 3.2].

9.11. Proposition. Let A be a C*-algebra with continuous trace. If A has a countable approximate identity, then every derivation of $\Gamma(K)$ is inner.

Proof. The first part of the proof proceeds as in [1, Theorem 3.2]. Since $\hat{A}$ is σ-compact it is paracompact. Thus by hypothesis there is an open cover $\{G_\lambda\}_{\lambda\in\Lambda}$ of $\hat{A}$ together with a family $\{e_\lambda\}_{\lambda\in\Lambda}$ in A^+ such that for each $\pi \in G_\lambda$ the operator $\pi(e_\lambda)$ is a one dimensional projection. We may assume that the covering is locally finite. Choose a partition of unity $\{f_\lambda\}_{\lambda\in\Lambda}$ subordinate to the covering $\{G_\lambda\}_{\lambda\in\Lambda}$.

Let $\pi \in \hat{A}$. By virtue of 7.3, the induced representation

$$\overline{\pi}:\Gamma(K) \to \Gamma(K_{\pi(A)})$$

is onto. Hence

$$\overline{\pi}(\Gamma(K)) = B(H_\pi).$$

Now let δ be a derivation of $\Gamma(K)$. By 9.2, $\overline{\pi} \cdot \delta$ is a derivation of $B(H_\pi)$. Just as in the proof of [1, Theorem 3.2] if $\pi \in G_\lambda$, let $x_\lambda(\pi)$ denote that unique operator in $B(H_\pi)$ which determines $\overline{\pi} \circ \delta$

(that is,

$$\overline{\pi} \circ \delta(\overline{\pi}(y)) = [x_\lambda(\pi), \overline{\pi}(y)])$$

and which satisfies

$$\pi(e_\lambda)x_\lambda(\pi)\pi(e_\lambda) = 0;$$

if $\pi \notin G_\lambda$, let $x_\lambda(\pi) = 0$. Set

$$x(\pi) = \Sigma f_\lambda(\pi)x_\lambda(\pi).$$

We will now show that for each compact subset F of $\hat{A}$ there is a z_F in $\Delta(K_A)$ such that $\pi(z_F) = x(\pi)$ for each $\pi \in F$. Let

$$J = \{a \in A : a|F = \{0\}\}$$

which is a closed two sided ideal of A. The C*-algebra A/J is also with continuous trace and has compact spectrum $\widehat{A/J} = F$. Let $\phi : A \to A/J$ be natural mapping. By 7.3, the induced map

$$\overline{\phi} : \Gamma(K) \to \Gamma(K_{\phi(A)})$$

is onto and by 10.13

$$\Gamma(K_{\phi(A)}) = \Delta(K_{\phi(A)}) = M(A/J).$$

By 9.2, $\overline{\phi} \circ \delta$ is a derivation of M(A/J). But from the proof of [1, Theorem 3.2] we see that the element x_F defined by $x_F(\pi) = x(\pi)$ for each $\pi \in F$ belongs to M(A/J) and

$$\overline{\phi} \circ \delta(y) = [x_F, y],\ y \in M(A/J).$$

Since $\overline{\Phi}$ maps $\Delta(K_A)$ onto $M(A/J)$ by 7.1, there is a z_F in $\Delta(K_A)$ such that $\pi(z_F) = x(\pi)$ for each $\pi \in F$.

Now consider the net $\{z_F\}_{F \in \mathfrak{F}}$, where $\mathfrak{F}$ denotes the family of all compact subsets of $\hat{A}$. Here $F_1 \leq F_2$ if $F_1 \subseteq F_2$. Let $a \in K_A$. By 5.39,

$$\{\pi \in \hat{A} : \|\pi(a)\| > 0\}$$

is contained in a compact subset F_0 of $\hat{A}$. It follows that for $F_1, F_2 \geq F_0$ we have

$$\|(z_{F_1} - z_{F_2})a\| = \sup\{\|\pi(z_{F_1} - z_{F_2})\pi(a)\| : \pi \in \hat{A}\}$$

$$= \sup\{\|\pi(z_{F_1} - z_{F_2})\pi(a)\| : \pi \in F_0\} = 0.$$

Similarly,

$$\|a(z_{F_1} - z_{F_2})\| = 0.$$

Therefore, $\{z_F\}_{F \in \mathfrak{F}}$ is κ-Cauchy and consequently, converges to some $z \in \Gamma(K_A)$. Clearly, $\pi(z) = x(\pi)$ for each $\pi \in \hat{A}$ and $\delta(y) = [z,y]$. Hence, our proof is complete.

CHAPTER 10. PCS-ALGEBRAS

Let A be a C*-algebra and let K_A denote its Pedersen ideal. The phenomenon $\Delta(K_A) = \Gamma(K_A)$ occurs frequently. In fact, this equality holds whenever A is commutative and has pseudocompact spectrum, or whenever A is élémentaire. In this chapter, we will study those C*-algebras A for which $\Delta(K_A) = \Gamma(K_A)$.

10.1. Theorem. Let A be a C*-algebra and let K_A denote its Pedersen ideal. Then the following statements are equivalent:

(i) $\Delta(K_A) = \Gamma(K_A)$;

(ii) A is a left ideal in $\Gamma(K_A)$;

(iii) the κ-bounded subsets of A are uniformly bounded;

(iv) for every sequence $\{x_n\}$ of elements in K_A with the property $\|x_n\| \to \infty$, the sequence of partial sums $\{\Sigma_{k=1}^n x_k\}$ is not κ-Cauchy;

(v) for every sequence $\{x_n\}$ of elements in K_A^+ with the property $\|x_n\| \to \infty$ and $x_i x_j = 0$, $i \neq j$, the sequence of partial sums $\{\Sigma_{k=1}^n x_k\}$ is not κ-Cauchy.

Proof. Clearly (i) implies (ii), so assume (ii) holds and that M is a κ-bounded subset of A that is not uniformly bounded. Clearly, we can choose a sequence $\{z_n\}$ of elements in M such that

$$\|z_n\| \geq 16^n.$$

Now set

$$x_n = z_n^* z_n / \| z_n \|^{3/2}$$

and note that $x_n \geq 0$ and

$$\| x_n \| \geq 4^n.$$

Let $a \in K_A^+$. Since M is κ-bounded, there is a $\delta > 0$ such that

$$\| z_n a \| < \delta$$

for each integer n. It follows that

$$\| (\Sigma_{k=m}^n x_k) a \| \leq \Sigma_{k=m}^n \| z_k^* z_k a \| / \| z_k \|^{3/2}$$

$$\leq \delta \Sigma_{k=m}^n 1/\| z_k \|^{1/2} \leq \delta \Sigma_{k=m}^n 1/4^k \leq \delta / 4^{m-1}.$$

Consequently, the sequence of partial sums $\{\Sigma_{k=1}^n x_k\}$ is κ-Cauchy. Since $\Gamma(K_A)$ is complete, the sequence of partial sums $\{\Sigma_{k=1}^n x_k\}$ converges in the κ-topology to some element x in $\Gamma(K_A)$. Now let

$$y_n = x_n / \| x_n \|^{3/2}.$$

Clearly,

$$\| y_n \| \leq 1/2^n.$$

Therefore, the sequence of partial sums $\{\Sigma_{k=1}^n y_k\}$ converges uniformly to some element y in A^+. Since A is a left ideal in $\Gamma(K_A)$,

$$\| y^{1/2} x y^{1/2} \|$$

is well defined, so

$$\|y^{1/2}xy^{1/2}\| = \|\Sigma_{k=1}^{\infty} y^{1/2}x_k y^{1/2}\| \geq \|y^{1/2}x_n y^{1/2}\|$$

$$= \|x_n^{1/2}yx_n^{1/2}\| = \|\Sigma_{k=1}^{\infty} x_n^{1/2}y_k x_n^{1/2}\| \geq \|x_n^{1/2}y_n x_n^{1/2}\|$$

$$= \|x_n^2\|/\|x_n\|^{3/2} = \|x_n\|^{1/2} \geq 2^n \to \infty,$$

which is a contradiction. Hence, (ii) implies (iii).

By a contradiction argument it is straightforward to show that (iii) implies (iv) and (iv) implies (v) is clear. Therefore, to complete the proof we need to show that (v) implies (i). To this end we will show that if

$$\Delta(K_A) \neq \Gamma(K_A),$$

then there exists a sequence $\{x_n\}$ of elements in K_A^+ with the property

$$\|x_n\| \to \infty$$

and $x_i x_j = 0$, $i \neq j$, and such that the sequence of partial sums $\{\Sigma_{k=1}^n x_k\}$ is κ-Cauchy.

Suppose

$$\Delta(K_A) \neq \Gamma(K_A)$$

and that $\{d_\lambda\}$ is a positive approximate identity for A contained in K_A with

$$\|d_\lambda\| \leq 1.$$

Since there is an element x in $\Gamma(K)$ such that

$$\|xd_\lambda\| \to \infty,$$

we can find a sequence $\{d_{\lambda_k}\}$ of elements in K_A such that the following hold:

(1) $\|d_{\lambda_{n+1}} d_{\lambda_k} - d_{\lambda_k}\| < 1/n$ for $k = 1,2,\ldots,n$;

(2) $\|xd_{\lambda_n}\| \to \infty$.

Let

$$A_0 = \{z \in A \mid \lim zd_{\lambda_n} = \lim d_{\lambda_n} z = z\}.$$

It is clear that A_0 is a closed self-adjoint subalgebra of A that contains $\{d_{\lambda_n}\}$ as an approximate identity. Since A_0 has a countable approximate identity, we have by [37, Proposition 3.1, p. 480] that A_0 has a positive approximate identity $\{e_n\}_{n=1}^{\infty}$ such that

$$e_{n+1}e_n = e_n \text{ and } \|e_n\| \leq 1$$

for each positive integer n. Note that $\{e_n\} \subset K_{00}^+$. By virtue of 3.4,

$$\liminf_n \|xe_n\| \geq \liminf_n \|xe_n d_{\lambda_k}\| = \lim_n \|x(e_n d_{\lambda_k})\| = \|xd_{\lambda_k}\|.$$

Since

$$\|xd_{\lambda_k}\| \to \infty,$$

we may assume that

$$\|xe_{2n+1}\| \geq 2^{n+1} + \|xe_{2n}\|.$$

Now set

$$y_n = (e_{2n} - e_{2n-1})x^*x(e_{2n} - e_{2n-1})$$

and

$$x_n = y_n/2^n.$$

Note that

$$\|y_n\| = \|xe_{2n} - xe_{2n-1}\|^2 \geq (\|xe_{2n}\| - \|xe_{2n-1}\|)^2 \geq 4^n.$$

It is clear that

$$x_n \in K_A^+, \|x_n\| \to \infty, \text{ and } x_i x_j = 0, i \neq j.$$

All that remains to be shown is that the sequence $\{\Sigma_{k=1}^n x_k\}$ is κ-Cauchy. Let

$$a \in K_A, \|a\| \leq 1,$$

and $\epsilon > 0$. Since we have by 3.4 that the multiplication on the left by x^*x is bounded on $\mathcal{L}_a$, there exists a positive integer N such that

$$\Sigma_{k=m}^n 1/2^k < \epsilon/\|L_{x^*x}|\mathcal{L}_a\|$$

for $n > m \geq N$. Here

$$L_{x^*x}(y) = x^*xy$$

for $y \in \Gamma(K)$. It follows that

$$\|\Sigma_{k=m}^n x_k a\| = \|\Sigma_{k=m}^n (1/2^k)(e_{2k} - e_{2k-1})L_{x^*x}((e_{2k} - e_{2k-1})a\|$$

$$\leq \Sigma_{k=m}^n \|L_{x^*x}|\mathcal{L}_a\|/2^k < \epsilon$$

for $n > m \geq N$. Clearly

$$\|a\, \Sigma_{k=m}^n x_k\| < \epsilon.$$

Hence the sequence of partial sums $\{\Sigma_{k=1}^{n} x_k\}$ is κ-Cauchy and our proof is complete.

10.2. Definition. Let A be a C*-algebra. We will call A a PCS-algebra if A satisfies any of the equivalent conditions of 10.1.

10.3. Corollary. Let A be a C*-algebra of B(H). If $K_A[H] = H$, then A is a PCS-algebra.

Proof. See 4.4.

Remark. There exist non-degenerate PCS-algebras of operators for which $K_A[H] \neq H$. Let $R = (-\infty,\infty)$, $N = \{1,2,3,\dots\}$ and $\Lambda = \beta R \setminus (\beta N \setminus N)$. Then Λ is pseudocompact [17, p. 97]. Let $A = C_0(\Lambda)$ and view it as an algebra of operators on $\ell_2(\Lambda)$ in the usual way. Since N is a closed non-compact subset of Λ, it is straightforward to see that $K_A[H] \neq H$.

10.4. Corollary. Let A be a C*-algebra and π an irreducible representation of A. Then $\pi(A)$ is a PCS-algebra.

Proof. The proof follows immediately from 10.3 and [13, 2.8.3, p. 44].

10.5. Corollary. Let A be a C*-algebra and I a primitive ideal in A. Then the quotient algebra A/I is a PCS-algebra.

10.6. Lemma. Let A be a C*-algebra and $\pi_1,\pi_2,\dots,\pi_n$ irreducible representations of A onto $H_1,H_2,\dots,H_n$, respectively. Let $r > 0$ and for each integer i, $1 \leq i \leq n$, let h_i, h_i' be elements in H_i so that

$$\|h_i'\| = 1 \text{ and } \|h_i\| \leq r.$$

Then there exists an element x in A,

$$\|x\| \leq 4rn^{1/2},$$

such that

$$\pi_i(x)(h_i') = h_i,\ i = 1,2,\ldots n.$$

Proof. This result is a trivial variant of [13, 2.8.2, p. 43] and [13, 2.8.3, p. 44].

10.7. Theorem. Let A be a C*-algebra and let $\{x_n\}$ be an orthogonal sequence in K_A^+ (that is, $x_n x_m = 0$, $n \neq m$) such that the sequence of partial sums $\{\Sigma_{k=1}^n x_k\}$ is κ-Cauchy. Let $a \in K_A$ and let $\{\alpha_n\}$ be the sequence defined by

$$\alpha_n = \sup\ \{\|\pi(a)\| \,|\, \pi \in \hat{A} \text{ and } \|\pi(x_n)\| > \|x_n\|/2\},$$

where $\hat{A}$ denotes the spectrum of A. If

$$\|x_n\| \to \infty,$$

then $\alpha_n \to 0$.

Proof. Suppose

$$\|x_n\| \to \infty$$

and $\{\alpha_n\}$ does not converge to 0. Since any subsequence of $\{x_n\}$ also satisfies the conditions of our hypothesis, we may assume without loss of generality that $\{\alpha_n\}$ is bounded away from 0, that is, there exists a $\delta > 0$ such that $\alpha_n > \delta$ for each positive integer n; moreover, we may assume that

$$(*)\quad \|x_n\| \geq n4^n.$$

It follows that we can find a sequence $\{\pi_n\}$ in $\hat{A}$ such that

$$\|\pi_n(x_n)\| > \|x_n\|/2 \text{ and } \|\pi_n(a)\| > \delta\ .$$

Now choose an $h \in H_{\pi_1}$ so that

$$\|h\| = 1 \text{ and } \|\pi_1(x_1)(h)\| > \|x_1\|/2,$$

where H_{π_1} is the associated Hilbert space of the representation π_1. Since

$$\|\pi_1(a)\| > \delta,$$

there is an $h' \in H_{\pi_1}$ such that

$$\|h'\| = 1 \text{ and } \|\pi_1(a)(h')\| > \delta.$$

It follows from 10.6 that there exists a $y \in A$ such that

$$h = \pi_1(ya)(h') \text{ and } \|y\| \leq 4/\delta.$$

We will now define by induction a sequence of quadruples $\{(n_k,y_k,h_k,h_k')\}_{k=1}^{\infty}$ that satisfies the following:

(a) $\{n_k\}$ is a subsequence of the positive integers;

(b) h_k, h_k' are elements in $H_{\pi_{n_k}}$ such that

$$\|h_k\| = \|h_k'\| = 1, \; \|\pi_{n_k}(x_{n_k})(h_k)\| > \|x_{n_k}\|/2,$$

$$\|\pi_{n_k}(a)(h_k')\| > \delta;$$

(c) $y_k \in A$, $\|y_k\| \leq 4k^{1/2}/\delta$, $\pi_{n_i}(y_k a)(h_i') = 0$ for $i = 1,2,\ldots,k-1$, and

$$\pi_{n_k}(y_k a)(h_k') = h_k;$$

(d) $\|\Sigma_{j=1}^{k-1} \|x_{n_j}\|^{-1/2} x_{n_k} y_j a\| \leq 1.$

Set $n_1 = 1$, $y_1 = y$, $h_1 = h$, and $h_1' = h'$. Suppose $\{(n_i,y_i,h_i,h_i')\}_{i=1}^{k-1}$ has already been defined. Since $\{\Sigma_{i=1}^{n} x_i\}$ is κ-Cauchy,

$$\|x_n b\| \to 0$$

for every $b \in K_A$. Consequently, there is a positive integer $n_k > n_{k-1}$ such that

$$\|\Sigma_{j=1}^{k-1} \|x_{n_j}\|^{-1/2} x_{n_k} y_j a\| \le 1.$$

Since

$$\|\pi_{n_k}(x_{n_k})\| > \|x_{n_k}\|/2$$

and

$$\|\pi_{n_k}(a)\| > \delta,$$

there exist elements h_k, h_k in $H_{\pi_{n_k}}$ such that

$$\|h_k\| = \|h_k'\| = 1,$$

$$\|\pi_{n_k}(x_{n_k})(h_k)\| > \|x_{n_k}\|/2,$$

and

$$\|\pi_{n_k}(a)(h_k')\| > \delta.$$

By 10.6, there is a $y_k \in H_{\pi_k}$,

$$\|y_k\| \le 4k^{1/2}/\delta,$$

such that

$$\pi_{n_i}(y_k a)(h_i') = 0$$

for $i = 1,2,\ldots,k-1$ and

$$\pi_{n_k}(y_k a)(h_k') = h_k.$$

Hence our induction is complete.

By (*) and (c)

$$\|y_k\| / \|x_{n_k}\|^{1/2} \leq (4\delta)(1/2^k).$$

Consequently, the sequence of partial sums

$$\{\Sigma_{k=1}^{p} (1/\|x_{n_k}\|^{1/2}) y_k)$$

is uniformly Cauchy and therefore has a limit z in A. Since $za \in K_A$ and $\{\Sigma_{k=1}^{n} x_k\}$ is κ-Cauchy, it follows that $\|x_n za\| \to 0$. However, by (b), (c), and (d),

$$\|x_{n_k} za\| \geq \|\pi_{n_k}(x_{n_k} za)(h_k')\| \geq \|(1/\|x_{n_k}\|^{1/2})\pi_{n_k}(x_{n_k} y_k a)(h_k')$$

$$+ \Sigma_{j=1}^{k-1} (1/\|x_{n_j}\|^{1/2})\pi_{n_k}(x_{n_k} y_j a)(h_k')\|$$

$$\geq \|\pi_{n_k}(x_{n_k})(h_k)\| / \|x_{n_k}\|^{1/2} - 1$$

$$\geq \|x_{n_k}\|^{1/2}/2 - 1.$$

Since

$$\|x_{n_k}\|^{1/2} \to \infty,$$

we have a contradiction and our proof is complete.

10.8. <u>Corollary</u>. If A is a C^*-algebra with compact spectrum $\hat{A}$, then A is a PCS-algebra.

Proof. Let $\pi \in A$. Since K_A^+ is dense in A^+, there exists an element a_π in K_A^+ such that

$$\|\pi(a_\pi)\| > 1.$$

Now set

$$V_\pi = \{\rho \in \hat{A} \mid \|\rho(a_\pi)\| > 1\}.$$

Due to the fact that the map

$$\rho \to \|\rho(a_\pi)\|$$

is lower semi-continuous, V_π is an open neighborhood of π. It follows that the collection of open sets $\{V_\pi \mid \pi \in A\}$ is an open cover of A. Consequently, since A is compact, there exists a finite subcover $V_{\pi_1}, V_{\pi_2}, \ldots, V_{\pi_n}$. Set

$$a = \Sigma_{i=1}^{n} a_{\pi_i}.$$

Clearly, $a \in K_A$ and

$$\|\pi(a)\| > 1$$

for each $\pi \in A$. If A is not a PCS-algebra, then a contradiction follows immediately from 10.1 and 10.7. Hence our proof is complete.

10.9. Corollary. Suppose A is the minimal non-zero closed two sided ideal of the factor B. Then A is a PCS-algebra; consequently, $\Gamma(K_A) = B$.

Proof. Since A is topologically simple, the first assertion follows from 10.8. The second assertion follows from [37, Corollary 2.2, p. 478].

In the next lemma let A be a C*-algebra and I a closed two sided ideal in A whose spectrum $\hat{I}$ is Hausdorff and dense in $\hat{A}$.

10.10. Lemma. If A is not a PCS-algebra, then there exists a sequence $\{x_n\}$ in K_I^+ such that the following hold:

$$\|x_n\| \to \infty;$$

the functions on $\hat{I}$ defined by

$$\pi \to \|\pi(x_n)\|$$

have pairwise disjoint compact supports; the sequence of partial sums $\{\Sigma_{k=1}^n x_k\}$ is κ_A-Cauchy.

Proof. Suppose A is not a PCS-algebra. By 10.1 there is a sequence $\{y_n\}$ in K_A^+ such that

$$\|y_n\| \to \infty,$$

$y_n y_m = 0$ for $n \neq m$, and the sequence of partial sums $\{\Sigma_{k=1}^n y_k\}$ is κ_A-Cauchy. Since the function

$$\pi \to \|\pi(y_n)\|$$

is lower semi-continuous on $\hat{A}$, the set

$$\{\pi \in A | \|\pi(y_n)\| > \|y_n\|/2\}$$

is a non-empty open subset of $\hat{A}$. Due to the fact that $\hat{I}$ is dense in $\hat{A}$, there exists, for each positive integer n, a $\pi_n \in \hat{I}$ such that

$$\|\pi_n(y_n)\| > \|y_n\|/2.$$

Due to the fact that for any subsequence $\{y_{n_k}\}$ the sequence of partial sums $\{\Sigma_{k=1}^{p} y_{n_k}\}$ is also κ_A-Cauchy, we can assume by virtue of 10.4 that the elements of the sequence $\{\pi_n\}$ are distinct. Since $\hat{I}$ is a locally compact Hausdorff space [13, 3.3.8, p. 64], we can find, by a straightforward topological argument, a subsequence $\{\pi_{n_k}\}$ and a sequence $\{V_k\}$ of pairwise disjoint compact neighborhoods in $\hat{I}$ so that π_{n_k} belongs to the interior of V_k. Now choose a positive increasing approximate identity $\{e_\lambda\}$ for I that is contained in K_I and

$$\|e_\lambda\| \leq 1$$

for each λ. Since each π_{n_k} is irreducible, there exists an e_{n_k} in K_I^+ such that

$$\|\pi_{n_k}((y_{n_k} e_{\lambda_k} y_{n_k})^{1/2})\| \geq \|y_{n_k}\|/4.$$

Next choose for each integer k a continuous function f_k on $\hat{I}$ with values [0,1] such that

$$f_k(\pi_{n_k}) = 1$$

and f_k is equal to 0 on the complement of V_k. Since K_I is a two sided ideal in A and K_I^+ contains its square roots, we have

$$(y_{n_k} e_{\lambda_k} y_{n_k})^{1/2}$$

in K_I^+. Moreover, by virtue of 5.40 there is an element x_k in K_I^+ such that

$$\pi(x_k) = f_k(\pi)\pi((y_{n_k} e_{\lambda_k} y_{n_k})^{1/2}))$$

for each $\pi \in \hat{I}$. Clearly, the functions on $\hat{I}$ defined by

$$\pi \to \|\pi(x_k)\|$$

have pairwise disjoint compact supports. Since

$$\|x_k\| \geq \|\pi_{n_k}(x_k)\| = \|\pi_{n_k}((y_{n_k} e_{\lambda_k} y_{n_k})^{1/2})\| \geq \|y_{n_k}\|/4,$$

we have

$$\|x_k\| \to \infty.$$

Therefore, it remains to be shown that the sequence of partial sums $\{\Sigma_{k=1}^n x_k\}$ is κ_A-Cauchy. If $a \in K_A^+$, then

$$\|(\Sigma_{k=p}^q x_k)a\|^2 = \|\Sigma_{k=p}^q ax_k^2 a\| \leq \|\Sigma_{k=p}^q ay_{n_k} e_{\lambda_k} y_{n_k} a\|$$

$$\leq \|\Sigma_{k=p}^q ay_{n_k}^2 a\| \leq \|\Sigma_{k=n_p}^{n_q} ay_k^2 a\| = \|(\Sigma_{k=n_p}^{n_q} y_k)a\|^2.$$

Since the sequence of partial sums $\{\Sigma_{k=1}^n y_k\}$ is κ_A-Cauchy, $\{\Sigma_{k=1}^n x_k\}$ is κ_A-Cauchy. Hence our proof is complete.

A point t in a topological space T is called a Hausdorff point if given any point t' in T, not in the closure of $\{t\}$, there exists disjoint neighborhoods of t and t'. We shall let S(T) represent the set of closed, Hausdorff points of T, and $S^0(T)$ the interior of S(T).

10.11. <u>Theorem</u>. Suppose A is a C*-algebra such that $S^0(\hat{A})$ is dense in the spectrum $\hat{A}$. If $\hat{A}$ is pseudocompact, then A is a PCS-algebra.

<u>Proof</u>. Suppose $\hat{A}$ is pseudocompact and A is not a PCS-algebra. Let I be the closed two sided ideal in A whose spectrum $\hat{I} = S^0(\hat{A})$. By 10.10, there exists a sequence $\{x_n\}$ in K_I^+ such that the following hold:

$$\|x_n\| \to \infty;$$

the functions on $\hat{I}$ defined by

$$\pi \to \|\pi(x_n)\|$$

have pairwise disjoint compact supports; the sequence of partial sums $\{\Sigma_{k=1}^{n} x_k\}$ is κ_A-Cauchy. Let

$$V_n = \{\pi \in \hat{I} \mid \|\pi(x_n)\| \geq \|x_n\|/2\}.$$

By virtue of [13, 3.3.2, p. 63], [13, 3.3.6, p. 64], and [13, 3.3.7, p. 64], and [15, 3.2, p. 51] the collection $\{V_n\}$ is a sequence of non-empty pairwise disjoint compact neighborhoods in $S^O(\hat{A})$ that are closed in $\hat{A}$. Furthermore, $\cup V_n$ is not closed in $\hat{A}$. For suppose $\cup V_n$ is closed. Then for each set V_n we can find a real-valued continuous function f_n on $\hat{A}$ such that $\|f_n\| = n$ and f_n vanishes off V_n; consequently, Σf_n is continuous on A contradicting the fact that $\hat{A}$ is pseudocompact.

Now let π_0 be a cluster point of $\cup V_n$ and suppose $\pi_0 \notin \cup V_n$. Since K_A^+ is dense in A^+, there exists an $a \in K_A^+$ such that

$$\|\pi_0(a)\| > 0.$$

So set

$$U = \{\pi \in \hat{A} \mid \|\pi(a)\| > \|\pi_0(a)\|/2\},$$

which is an open neighborhood of π_0, since the function on $\hat{A}$ defined by

$$\pi \to \|\pi(a)\|$$

is lower semi-continuous. By a straightforward compactness argument, we can show that π_0 and each set V_n can be separated, consequently, there is a subsequence $\{V_{n_k}\}$ such that $U \cap V_{n_k}^O \neq \phi$, where $V_{n_k}^O$ denotes the interior of V_{n_k}. But this contradicts 10.7. Hence A must be a PCS-algebra and our proof is complete.

10.12. Corollary. Let A be a separable C*-algebra such that each irreducible representation of A is finite dimensional. If the spectrum of A is pseudocompact, then A is a PCS-algebra.

Proof. The proof follows immediately from [14, Proposition 2, p. 117] and 10.11.

10.13. Corollary. Let A be a C*-algebra with continuous trace. If the spectrum of A is pseudocompact, then A is a PCS-algebra.

Proof. The proof follows from [13, 4.5.3, p. 93] and 10.11.

10.14. Theorem. Let A be a C*-algebra. If A is a PCS-algebra, then $\hat{A}$ is pseudocompact; moreover, if $\hat{A}$ is Hausdorff and pseudocompact, then A is a PCS-algebra.

Proof. Let Z be the center of $\Gamma(K_A)$. For each $z \in Z$ let f_z be the complex valued continuous function on $\hat{A}$ defined as in 5.42. The map $z \to f_z$ is a *-isomorphism of Z onto $C(\hat{A})$. So if A is a PCS-algebra, then

$$\sup_{\pi \in \hat{A}} |f_z(\pi)| = \sup_{\pi \in \hat{A}} \|\pi(z)\| \leq \|z\| < \infty.$$

Hence, $\hat{A}$ is pseudocompact. If $\hat{A}$ is Hausdorff and pseudocompact, then A is a PCS-algebra by virtue of 10.11.

Question. Can the Hausdorff condition of 10.14 be dropped?

The next result is a generalization of a theorem of R. S. Phillips [34] on the nonexistence of projections of ℓ^∞ onto c_0. A commutative version of this result is due to J. B. Conway [8].

10.15. Theorem. Let A be a C*-algebra, $\{e_\lambda\}$ an approximate identity for A, and B a C*-subalgebra of A that contains $\{e_\lambda\}$. If A is complemented in $\Delta(K_A)$, then B is a PCS-algebra.

Proof. Suppose B is not a PCS-algebra. By virtue of 10.1 there exists a sequence $\{y_n\}$ in K_B^+ that satisfies the following:

$$\|y_n\| \to \infty;$$

$y_n y_m = 0$ for $n \neq m$; the sequence of partial sums $\{\Sigma_{k=1}^n y_k\}$ is κ_B-Cauchy. Now set

$$x_n = (1/\|y_n\|)y_n.$$

Since

$$\|\Sigma_{k=m}^n x_k a\|^2 = \|\Sigma_{k=m}^n a^* x_k^2 a\| \leq \|\Sigma_{k=m}^n a^* y_k^2 a\| = \|\Sigma_{k=m}^n y_k a\|^2$$

for each $a \in K_B$, it follows that the sequence of partial sums $\{\Sigma_{k=1}^n x_k\}$ is also κ_B-Cauchy. Note that

$$\|\Sigma_{k=m}^n x_k\| = 1.$$

We now wish to show that the sequence of partial sums $\{\Sigma_{k=1}^n x_k\}$ is κ_A-Cauchy. Let $a \in K_A^+$ and $\epsilon > 0$. Since $e_\lambda a \to a$, we can find an e_λ so that

$$\|e_\lambda a - a\| < \epsilon/2.$$

Since any approximate identity for B will be an approximate identity for A, we may assume that $\{e_\lambda\}$ is positive, increasing, and contained in K_B. Because $\{\Sigma_{k=1}^n x_k\}$ is κ_B-Cauchy, it follows that there is a positive integer N such that

$$\|\Sigma_{k=m}^n x_k e_\lambda\| < \epsilon/2\|a\|$$

for $n \geq m > N$. Consequently,

$$\|\Sigma_{k=m}^{n} x_k a\| \leq \|\Sigma_{k=m}^{n} x_k e_\lambda a\| + \|\Sigma_{k=m}^{n} x_k(e_\lambda a - a)\|$$

$$\leq \|\Sigma_{k=m}^{n} x_k e_\lambda\| \|a\| + \|e_\lambda a - a\| < \epsilon$$

for $m \geq n > N$, so $\{\Sigma_{k=1}^{n} x_k\}$ is κ_A-Cauchy and therefore has a κ_A-limit in the unit ball of $\Delta(K_A)$, which we denote by $\Sigma_{k=1}^{\infty} x_k$. Similarly, for $\alpha = \{\alpha_k\} \in \ell^\infty$, the sequence of partial sums $\{\Sigma_{k=1}^{n} \alpha_k x_k\}$ has a κ_A-limit $\Sigma_{k=1}^{\infty} \alpha_k x_k$ in $\Delta(K_A, \| \|)$ of norm $\|\alpha\|_\infty$.

Define the linear map $T_1 : \ell^\infty \to \Delta(K_A)$ by the formula

$$T_1(\alpha) = \Sigma_{k=1}^{\infty} \alpha_k x_k,$$

where $\alpha = \{\alpha_k\}$. We have already established that T_1 is an isometry. Now let B_0 denote the commutative C*-algebra generated by $x_1, x_2, \ldots$. By the Gelfand mapping theorem we can view B_0 as $C_0(\hat{B}_0)$. Since

$$\|x_n\|_\infty = 1,$$

we can pick a sequence $\{t_n\}$ in $\hat{B}_0$ such that $x_n(t_n) = 1$ for each n. By virtue of the Hahn-Banach theorem there exists a sequence $\{g_k\}$ of bounded linear functionals on A such that the following hold:

(i) $\|g_k\| = 1$;

(ii) $g_k(x) = \hat{x}(t_k)$ for each $x \in B_0$.

Now set $f_k = x_k \cdot g_k$, where

$$(x_k \cdot g_k)(a) = g_k(a x_k)$$

for each $a \in A$. Notice

$$\|f_k\| \leq 1.$$

Since the sequence of partial sums $\{\Sigma_{k=1}^{n} x_k\}$ is κ_A-Cauchy and uniformly bounded by 1,

$$\|ax_k\| \to 0$$

for each $a \in A$. Hence $\{f_k(a)\} \in c_0$ for each $a \in A$. Define the linear map $T_2 : A \to c_0$ by the formula $T_2(a) = \{f_k(a)\}$. Clearly, T_2 is bounded by 1. If A is complemented in $\Delta(K_A)$, then there exists a bounded projection P of $\Delta(K_A)$ onto A. Therefore, the map $\tau : \ell^\infty \to c_0$ defined by

$$T(\alpha) = T_2(P(T_1(\alpha)))$$

is bounded. Moreover, if

$$\alpha = \{\alpha_k\} \in c_0,$$

then

$$\tau(\alpha) = T_2(P(\Sigma_{k=1}^{\infty} \alpha_k x_k)) = T_2(\Sigma_{k=1}^{\infty} \alpha_k P(x_k)) = T_2(\Sigma_{k=1}^{\infty} \alpha_k x_k)$$

$$= \{f_n(\Sigma_{k=1}^{\infty} \alpha_k x_k)\} = \{\Sigma_{k=1}^{\infty} \alpha_k g_n(x_k x_n)\} = \{\alpha_n g_n(x_n^2)\}$$

$$= \{\alpha_n \hat{x}_n^2(t_n)\} = \{\alpha_n\} = \alpha.$$

This contradicts Phillips' theorem. Therefore B must be a PCS-algebra and our proof is complete.

BIBLIOGRAPHY

1. C. Akemann, G. Elliott, G. Pedersen, and J. Tomiyama, "Derivations and multipliers of C*-algebras", (to appear).

2. C. Akemann and G. Pedersen, "Complications of semicontinuity in C*-algebra theory", (to appear).

3. C. Akemann, G. Pedersen, and J. Tomiyama, "Multipliers of C*-algebras", J. Functional Analysis 13 (1973), 277-301.

4. M. Altman, "Factorisation dans les algebres de Banach", C. R. Acad. Sc. Paris t.272 (1971), 1388-1389.

5. R. Busby, "Double centralizers and extensions of C*-algebras", Trans. Amer. Math. Soc. 132 (1968), 79-99.

6. R. Busby, "On structure spaces and extensions of C*-algebras", J. Functional Analysis 1 (1967), 370-377.

7. F. Combes, "Poids sur une C*-algebra", J. Math. Pures et appl. 47 (1968), 57-100.

8. J. Conway, "Projections and retractions", Proc. Amer. Math. Soc. 17 (1966), 843-847.

9. J. Dauns, "Multiplier rings and primitive ideals", Trans. Amer. Math. Soc. 145 (1969), 125-158.

10. J. Dauns, "Spectral theory of algebras and adjunction of the identity", Math. Ann. 179 (1969), 175-202.

11. J. Dauns and K. Hofmann, "Representation of rings by sections", Mem. Amer. Math. Soc. 83 (1968), 1-180.

12. J. Dixmier, Les algebrés d'operatéurs dans l'espace Hilbertien, Deuxiéme Èdition, Gauthier-Villars, Paris, 1967.

13. J. Dixmier, Les C*-algèbres et leur représentations, Gauthier-Villars, Paris, 1964.

14. J. Dixmier, "Points separés dans le spectre d'une C*-algèbre", Acta Sci. Math. 22 (1961), 115-128.

15. J. Dixmier, "Traces sur les C*-algèbres II, Bull. Sci. Math. 88 (1964), 39-57.

16. J. Dixmier, "Ideal center of a C*-algebra", Duke Math. J. 35 (1968), 375-382.

17. L. Gillman and M. Jerison, "Rings of continuous functions", van Nostrand, Princeton, 1960.

18. E. Hewitt, "The range of certain convolution operators", Math. Scand. 15 (1964), 147-155.

19. K. Hofmann, "Gelfand-Naimark theorems for non-commutative topological rings", Proceedings of the Second Prague Topological Symposium, 1966, 184-189.

20. K. Hofmann, "Extending C*-algebras by adjoining an identity", Proceedings of the I. International Symposium on Extension Theory of Topological Structures and its Applications, held in Berlin 1967, 119-125.

21. B. Johnson, "An introduction to the theory of centralizers", Proc. London Math. Soc. 14 (1969), 299-320.

22. B. Johnson, "Centralizers on certain topological algebras", J. London Math. Soc. 39 (1964), 603-614.

23. B. Johnson and A. Sinclair, "Continuity of derivations and a problem of Kaplansky", Amer. J. Math. 90 (1968), 1067-1073.

24. A. Lazar and D. Taylor, "Double centralizers of Pedersen's ideal of a C*-algebra", Bull. Amer. Math. Soc. 78 (1972), 992-997.

25. A. Lazar and D. Taylor, "Double centralizers of Pedersen's ideal of a C*-algebra II", Bull. Amer. Math. Soc. 79 (1973), 361-366.

26. A. Lazar and D. Taylor, "A Dauns-Hofmann theorem for $\Gamma(K)$", Recent advances in the representation of algebras by continuous sections in sheaves and bundles, Memoirs Amer. Math. Soc. 148 (1974), 135-144.

27. G. Pedersen, "Measure theory for C*-algebras", Math. Scand. 19 (1966), 131-145.

28. G. Pedersen, "Measure theory for C*-algebras II", Math. Scand. 22 (1968), 63-74.

29. G. Pedersen, "A decomposition theorem for C*-algebras", Math. Scand. 22 (1968), 266-268.

30. G. Pedersen, "Measure theory for C*-algebras III", Math. Scand. 25 (1969), 71-93.

31. G. Pedersen, "Measure theory for C*-algebras IV", Math. Scand. 25 (1969), 121-127.

32. G. Pedersen, "Applications of weak* semicontinuity in C*-algebra theory", Duke Math. J. 39 (1972), 431-450.

33. G. Pedersen and N. Petersen, "Ideals in a C*-algebra", Math. Scand. 27 (1970), 193-204.

34. R. Phillips, "On linear transformations", Trans. Amer. Math. Soc. 48 (1940), 516-541.

35. S. Sakai, C*-algebras and W*-algebras, Springer-Verlag, New York, 1971.

36. D. Taylor, "Interpolation in algebras of operator fields", J. Functional Analysis 10 (1972), 159-190.

37. D. Taylor, "A general Phillips theorem for C*-algebras and some applications", Pacific J. Math. 40 (1972), 477-488.

Department of Mathematics, Tel-Aviv University, Tel-Aviv, Israel

Department of Mathematics, Montana State University, Bozeman, Montana

General instructions to authors for
PREPARING REPRODUCTION COPY FOR MEMOIRS

For more detailed instructions send for AMS booklet, "A Guide for Authors of Memoirs." Write to Editorial Offices, American Mathematical Society, P. O. Box 6248, Providence, R. I. 02940.

MEMOIRS are printed by photo-offset from camera copy fully prepared by the author. This means that, except for a reduction in size of 20 to 30%, the finished book will look exactly like the copy submitted. Thus the author will want to use a good quality typewriter with a new, medium-inked black ribbon, and submit clean copy on the appropriate model paper.

Model Paper, provided at no cost by the AMS, is paper marked with blue lines that confine the copy to the appropriate size. Auth[illegible] typewriter to be used has PICA-size (10 characters to the inch) or ELITE-size [illegible]

Line [illegible] For best appearance, and [illegible], a typewriter equipped with a half-space ratchet – 12 notches to the inch – shou[illegible]. (This may be purchased [illegible]ched at small cost.) Three notches make the desired spacing, which is equi[illegible] 1-1/2 ordinary single spaces. [illegible] copy has a great many subscripts and superscripts, however, double spacing shou[illegible].

Speci[illegible] may be filled in carefull[illegible]nd, using dense black ink, or INSTANT ("rub-on") LETTERING may be used. [illegible] has a sheet of several hundred [illegible]ed symbols and letters which may be purchased for $2.

Diagr[illegible] may be drawn in black ink either [illegible] on the model sheet, or on a separate sheet and pasted with rubber cement into s[illegible] text.

Page [illegible] in CAPITAL LETTERS (preferably), at the top of the page – just above the bl[illegible]

LEFT-hand, EVEN-numbered pages should be headed with the AUTHOR'S NAME;

RIGHT-hand, ODD-numbered pages should be headed with the TITLE of the paper (in shortened form if necessary). Page 1, of course, should have a display title instead of a running head, dropped 1 inch from the top blue line.

Page Numbers should be at the top of the page, on the same line with the running heads,

LEFT-hand, EVEN numbers – flush with left margin;

RIGHT-hand, ODD numbers – flush with right margin.

Exceptions – PAGE 1 should be numbered at BOTTOM of page, centered just below the footnotes, on blue line provided.

FRONT MATTER PAGES should also be numbered at BOTTOM of page, with Roman numerals (lower case), centered on blue line provided.

MEMOIRS FORMAT

It is suggested that the material be arranged in pages as indicated below.
Note: Starred items (*) are requirements of publication.

Front Matter (first pages in book, preceding main body of text).

Page i – *Title, *Author's name.

Page ii – *Abstract (at least 1 sentence and at most 300 words).

*AMS (MOS) subject classifications (1970). (These represent the primary and secondary subjects of the paper. For the classification scheme, see Appendix to MATHEMATICAL REVIEWS, Index to Volume 39, June 1970. See also June 1970 NOTICES for more details, as well as illustrative examples.)

Key words and phrases, if desired. (A list which covers the content of the paper adequately enough to be useful for an information retrieval system.)

Page iii – Table of contents.

Page iv, etc. – Preface, introduction, or any other matter not belonging in body of text.

Page 1 – *Title (dropped 1 inch from top line, and centered).

Beginning of Text.

Footnotes: *Received by the editor date.

Support information – grants, credits, etc.

Last Page (at bottom) – Author's affiliation.